W9-BHW-839

SAS® Programming Tips:

A Guide to Efficient SAS® Processing

Kut Zhang

SAS Institute Inc.
SAS Campus Drive
Cary, NC 27513

The correct bibliographic citation for this manual is as follows: SAS Institute Inc., *SAS® Programming Tips: A Guide to Efficient SAS® Processing,* Cary, NC: SAS Institute Inc., 1990. 155 pp.

SAS® Programming Tips: A Guide to Efficient SAS® Processing

SAS Institute Inc., SAS Campus Drive, Cary, North Carolina 27513.

1st printing, January 1991
2nd printing, April 1991
3rd printing, November 1992
4th printing, June 1994
5th printing, July 1996

Note that text corrections may have been made at each printing.

The SAS® System is an integrated system of software providing complete control over data access, management, analysis, and presentation. Base SAS software is the foundation of the SAS System. Products within the SAS System include SAS/ACCESS, SAS/AF, SAS/ASSIST, SAS/CALC, SAS/CONNECT, SAS/CPE, SAS/DMI, SAS/EIS, SAS/ENGLISH, SAS/ETS, SAS/FSP, SAS/GRAPH, SAS/IMAGE, SAS/IML, SAS/IMS-DL/I, SAS/INSIGHT, SAS/LAB, SAS/NVISION, SAS/OR, SAS/PH-Clinical, SAS/QC, SAS/REPLAY-CICS, SAS/SESSION, SAS/SHARE, SAS/STAT, SAS/TOOLKIT, SAS/TRADER, SAS/TUTOR, SAS/DB2, SAS/GEO, SAS/GIS, SAS/PH-Kinetics, SAS/SHARE*NET, SAS/SPECTRAVIEW, and SAS/SQL-DS software. Other SAS Institute products are SYSTEM 2000® Data Management Software, with basic SYSTEM 2000, CREATE, Multi-User, QueX, Screen Writer, and CICS interface software; InfoTap® software; NeoVisuals® software; JMP, JMP IN, and JMP Serve® software; SAS/RTERM® software; the SAS/C® Compiler and the SAS/CX® Compiler; Video Reality™ software; VisualSpace™ software; Budget Vision, CFO Vision, Compensation Vision, IT Service Vision, and Risk Vision™ software; Scalable Performance Data Server™ software; and Emulus® software. MultiVendor Architecture™ and MVA™ are trademarks of SAS Institute Inc. SAS Institute also offers SAS Consulting® and SAS Video Productions® services. *Authorline,* Books by Users, The Encore Series, *JMPer Cable, Observations, SAS Communications, SAS Professional Services, SAS Views,* the SASware Ballot, and SelecText, and Solutions@Work™ documentation are published by SAS Institute Inc. The SAS Video Productions logo, the Books by Users SAS Institute's Author Service logo, and The Encore Series logo are registered service marks or registered trademarks of SAS Institute Inc. The Helplus logo, the SelectText logo, the SAS Online Samples logo, the Video Reality logo, the Quality Partner logo, the SAS Business Solutions logo, and the SAS Rapid Warehousing Program logo are service marks or trademarks of SAS Institute Inc. All trademarks above are registered trademarks or trademarks of SAS Institute Inc. in the USA and other countries. ® indicates USA registration.

The Institute is a private company devoted to the support and further development of its software and related services.

Other brand and product names are registered trademarks or trademarks of their respective companies.

Doc S19, Ver 1.37N, 110990

Contents

Credits

Documentation

Composition	Gail C. Freeman, Cynthia M. Hopkins, Pamela A. Troutman, David S. Tyree
Graphic Design	Creative Services Department
Proofreading	Patsy P. Blessis, Carey H. Cox, Hanna P. Hicks, Josephine P. Pope, Toni P. Sherrill, Linda Rudd Wooten
Technical Review	Brian P. Bowman, Gloria N. Cappy, Tom Cole, Eloise M. Currie, Darylene C. Hecht, Linda C. Helwig, Brian Hess, Bernadette H. Johnson, Susan E. Johnston, Rusti Ludwick, Jean Moorefield, Mark Moorman, Susan M. O'Connor, Sally Painter, Randall D. Poindexter, Denise M. Poll, Dave Prinsloo, Lisa M. Ripperton, Richard G. Roach, Mark V. Schaffer, Douglas J. Sedlak, Maggie Underberg, Holly S. Whittle, Dea B. Zullo
Writing and Editing	Rick Early, Christina N. Harvey, Brenda C. Kalt, Len Olszewski, Helen Weeks, John M. West

Contributors

These individuals contributed tips or served as subject matter experts. Some of them also reviewed the book for technical accuracy.

Stephen Beatrous, David C. Berger, Jeff Cartier, Anne Corrigan, Ginny Dineley, Rajen H. Doshi, Lisa A. Horwitz, Paul M. Kent, Amy S. Kosarin, Lynn H. Patrick, Jeffrey A. Polzin, Meg Pounds, Joy Reel, Dan Squillace, Scott Sweetland, E. Tim Thompson, Bruce Tindall

Testing

The following persons supported our testing effort by writing our testing application, providing host operating system support, or testing the tips and compiling the results.

Jeff Cartier, Victoria Fritzler, Amy S. Glass, Selené Hudson, Ken Larsen, David Shubert, Joan M. Stout, David A. Teal

Acknowledgments

Many people made significant contributions to this book, either directly by responding to our requests for suggestions or indirectly by publishing a SUGI paper on efficient SAS programming. Those who contributed include

Marjorie Bedinger, Health Care Financing Administration
Jules Bosch, Bosch Systems
Eric Brinsfield, Meridian Software Inc.
Amy L. Caron, Syntex Laboratories
Tracy Cermack, American Honda Motor Co., Inc.
John Cohen, E. I. Du Pont de Nemours and Company
Ted Clay, Consultant
Bruce W. Densmore, Bank of Nova Scotia
Frank DiIorio, Computer Sciences Corporation
Akos Felsovalyi, Citibank
Donald J. Henderson, SAS Consulting Services Inc.
Neil Howard, ORI, Inc.
Sandra D. Iverson, Corporate Cost Management
Brett A. Knox, National Demographics
Dan E. Lodter, Electronic Data Systems
Wendy B. London, Pharmaceutical Research Associates, Inc.
Juliana M. Ma, Quintiles, Inc.
Myron Molnau, University of Idaho
Gary F. Plazyk, A. T. Kearney, Inc.
Merry Rabb, SAS Consulting Services Inc.
Clinton S. Rickards, Aetna Life Insurance Company
Kristen Rozenberg, Wyman-Gordon Company
David S. Rubin, Empire Blue Cross & Blue Shield
Howard Schreier, U. S. Department of Commerce
Stan Sibley, Sibley Enterprises
Susan J. Slaughter, Consultant
Bob L. Virgile, Consultant
Bucky Walsh, Pharmaceutical Research Associates, Inc.
Mark Watson, Consultant
Marianne Whitlock, Westat
Debbie D. Wilson, Commonwealth of Kentucky
Ann Yang, Consultant
James P. Young, Syntex Laboratories

SAS Institute appreciates the contribution of each of these individuals. The Publications Division encourages you to communicate your opinions and suggestions. It is through such communication that we hope to meet the needs of a wider audience of SAS users.

Using This Book

This preface describes *SAS Programming Tips: A Guide to Efficient SAS Processing* and tells you how to use the book most effectively.

Purpose

SAS Programming Tips: A Guide to Efficient SAS Processing is a handbook for improving the efficiency of your SAS applications. It suggests coding techniques, provides guidelines for their use, and compares examples of acceptable and improved ways to accomplish the same task. The tips are organized according to seven general principles of SAS programming, and each tip identifies the resource or resources improved by using the technique. Some explanation of internal SAS processing and general programming practices may be provided, but this book emphasizes practical examples and guidelines.

   SAS Programming Tips is not comprehensive in its treatment of efficiency issues. It concentrates on programming tasks common to most applications, and the topics are limited to the DATA step and the data summarization, utility, and reporting procedures of base SAS software in Release 6.06. This book does not address efficiencies in interactive processing or present techniques that require a knowledge of operating-system dependencies.

   SAS Programming Tips makes suggestions that are likely to improve performance on most operating systems. Because the factors determining the amount of improvement are too numerous, this book will not provide job statistics or attempt to quantify the techniques presented in this book. Instead, it suggests that you test performance within your operating system environment. To help in your testing, one chapter provides an elementary discussion of the SAS tools available for reporting job statistics.

Audience

SAS Programming Tips is written for experienced SAS programmers who must write applications under one or more of the following circumstances:

☐ Their computer resources are limited or costly.

☐ They develop applications for an end-user community.

☐ Their jobs run in a production environment.

☐ They process large data sets.

☐ They use the batch or the noninteractive method of running the SAS System.

   This book assumes readers have a working knowledge of the SAS System that is above the basic level. They can write SAS programs that work, and they have a strong interest in writing programs that use fewer human or computer resources.

Prerequisites

The following table summarizes the SAS System concepts you should understand to use *SAS Programming Tips*.

You should be able to	Refer to
invoke the SAS System at your site (this book assumes you use batch or noninteractive processing)	instructions provided by the SAS Software Consultant at your site
use base SAS software; you need varying amounts of familiarity with the SAS System depending on which features you want to use	□ *SAS Language and Procedures: Introduction, Version 6, First Edition* for a brief introduction
	□ *SAS Language and Procedures: Usage, Version 6, First Edition* for a more thorough introduction
	□ *SAS Language: Reference, Version 6, First Edition* for reference information for the SAS language
	□ *SAS Procedures Guide, Version 6, Third Edition* for usage and reference information for SAS procedures
manage files on your operating system	SAS documentation for your host operating system or vendor documentation for your host operating system

How to Use This Book

The following sections provide an overview of the information in this book, explain the organization, and describe how you can best use the book.

Organization

SAS Programming Tips is divided into four parts, and each part has one or more chapters that develop a topic.

Part 1: Understanding Efficiency

Part 1 defines efficiency in terms of human and computer resources and introduces the concept of tradeoffs in efficiency. Chapter 2 discusses the tools and procedures you can use to measure the performance of applications within your own computing system environment.

Chapter 1, "What Is Efficiency?"

Chapter 2, "How Do You Know if Your Program Is Efficient?"

Part 2: Tips for Efficient SAS Programming

Part 2 introduces the seven general principles of efficient SAS programming. Chapters 4 through 10 contain tips that show you how to apply the principles.

Chapter 3, "How Can Your SAS Programs Be More Efficient?"

Chapter 4, "Read and Write Data Selectively"

Chapter 5, "Execute Only the Statements You Need, in the Order You Need Them"

Chapter 6, "Take Advantage of SAS Procedures"

Chapter 7, "Know SAS System Defaults"

Chapter 8, "Control Sorting"

Chapter 9, "Test Your Programs, Know Your Data"

Chapter 10, "Code Clearly"

Part 3: Summary of Tips

Part 3 contains two chapters that summarize tips.

Chapter 11, "A Quick Look at Efficiency Tips"

Chapter 12, "Summary of Tips by Resource"

Part 4: Appendix

Part 4 contains one appendix.

Appendix, "Generating Data for Testing Program Efficiency"

What You Should Know About the Tips

The tips presented in this book are a compilation of programming techniques gathered from interested SAS System users. To verify that the techniques show some improvement in performance, the Publications Division tested those tips that save computer resources. Performance statistics for each pair of examples were collected and compared for the MVS, CMS, and VMS operating systems. Each example was submitted for five separate SAS sessions using SAS data sets with 1000 observations or raw data files with 1000 records. If a pair of examples appears in this book, it represents an improvement in performance on one or all operating systems under these test conditions.

The amount of savings, if any, you realize from using the tips in this book can vary with your computing environment and the context of your application. When you include the tips in this book in your applications, use the tools and techniques provided in Chapter 2 to verify the savings. You may need to try more than one tip or develop your own variation to achieve the most efficient results.

What You Should Read

First, read Part 1 for an overview of efficiency and program performance. Next, scan Part 3 to find the programming techniques that could be useful for your application. Refer to the tips in Part 2 for complete discussions.

Reference Aids

SAS Programming Tips will be easier to use if you are familiar with the following features of the book. They are listed in order of appearance within the book.

inside front cover graphic	illustrates the different parts of the SAS System, organized by function.
table of contents	lists the chapter titles, major subheadings, and the page numbers for each.
chapter tables of contents	give the titles and page numbers of sections within chapters. Chapter tables of contents appear in all chapters and in the appendix.
summary table and lists	provide an index and cross references to tips by resource saved.
glossary	defines the terms introduced in this book.
index	provides the page numbers where specific topics are discussed. Page ranges indicate discussions that cover several pages.
inside back cover graphic	illustrates the parts of base SAS software and the major categories of features within each.

Conventions

This section explains the conventions this book uses, including typographical conventions and conventions for referring to SAS data libraries and external files.

Typographical Conventions

SAS Programming Tips uses several type styles and related conventions in presenting information. The following list explains the meaning of the conventions in general.

roman	is the standard type style used for most text in this book.
UPPERCASE ROMAN	is used for literal elements of the SAS language and variable names in text.
italic	is used to define new terms, indicate glossary terms, and to emphasize important information.

`monospace`   is used to show examples of SAS code set off from the
                               text. In most cases, this book uses lowercase type for
                               SAS code, with the exception of some title characters.

`red monospace`   is used in examples of SAS code to highlight the SAS
                               statements that illustrate the tip.

The following examples illustrate these typographical conventions:

☐ An operator written with letters, such as EQ for =, is called a *mnemonic
   operator*.

☐ Use the TRIM function in the concatenation operation as follows:

```
data namegame;
   length color name $8;
   color='black';
   name='jack';
   game=trim(color)||name;
run;
```

This example produces a value of `blackjack` for the variable GAME.

Conventions for Referencing SAS Data Libraries and External Files

Many SAS statements and other elements of the SAS language refer to SAS data
libraries and external files. In Release 6.06, you can usually choose whether to
make the reference through a logical name (a libref or fileref) or to use the
physical filename enclosed in quotes. If you use a logical name, you usually have a
choice of using a SAS statement (LIBNAME or FILENAME) or the operating
system's control language to make the association. As a result, many methods of
referring to SAS data libraries and external files are available, and some of them
depend on the host operating system.

In examples that use external files, this book uses the italicized phrase
file-specification. You must see the SAS documentation for your host operating
system for the rules for referencing external files on your host. Similarly,
examples refer to SAS data libraries with the convention *SAS-data-library*. The
following example illustrates an INFILE statement that refers to an external file:

```
infile file-specification obs=100;
```

Additional Documentation

SAS Institute provides many publications about software products of the SAS System and how to use them on specific host operating systems. For a complete list of SAS publications, you should refer to the current *Publications Catalog*. The catalog is produced twice a year. You can order a free copy of the catalog by writing to

> SAS Institute Inc.
> Book Sales Department
> SAS Campus Drive
> Cary, NC 27513

Base SAS Software Documentation

You will find these other documents helpful when using base SAS software:

☐ *SAS Language and Procedures: Introduction, Version 6, First Edition* (order #A56074) provides information for users who are unfamiliar with the SAS System or any other programming language.

☐ *SAS Language: Reference, Version 6, First Edition* (order #A56076) provides detailed reference information about the elements of the DATA step in base SAS software.

☐ *SAS Procedures Guide, Version 6, Third Edition* (order #A56080) provides detailed reference information about procedures in base SAS software.

☐ *SAS Language and Procedures: Usage, Version 6, First Edition* (order #A56075) provides task-oriented examples of the major features of base SAS software.

☐ *SAS Guide to Macro Processing, Version 6, Second Edition* (order #A56041) provides a tool for extending and customizing your SAS programs.

☐ *SAS Guide to the SQL Procedure: Usage and Reference, Version 6, First Edition* (order #A56070) provides both tutorial and reference information on the SQL procedure.

☐ SAS documentation for your host operating system provides information about the host-specific features of the SAS System for your operating system.

Documentation for Other SAS Software

The SAS System includes many software products in addition to base SAS software. Several books that may be of particular interest to you are listed here:

□ *SAS/ASSIST Software: Your Interface to the SAS System, Version 6, First Edition* (order #A56086) provides information on using the SAS System in a menu-driven windowing environment that requires no programming.

□ *SAS/FSP Software: Usage and Reference, Version 6, First Edition* (order #A56001) provides information on using interactive procedures for creating SAS data sets and entering and editing data, and for creating, editing, and printing form letters and reports.

□ *SAS/AF Software: Usage and Reference, Version 6, First Edition* (order #A56011) provides tutorial and reference information about the applications development facilities available in SAS/AF software.

□ *SAS/GRAPH Software: Reference, Version 6, First Edition, Volume 1* and *Volume 2* (order #A56020) provides information on creating presentation graphics to illustrate relationships of data.

Part 1

Understanding Efficiency

Part 1 defines efficiency in terms of human and computer resources and introduces the concept of tradeoffs in efficiency. It also provides a discussion of the SAS System tools that are available for obtaining efficiency information.

Chapter **1** What Is Efficiency?

Introduction

The first step in improving efficiency is to understand what efficiency is. This chapter first defines efficiency as this book uses the term. Then it discusses when you should make your programs more efficient and how to do that. Finally, it shows some examples of how these concepts can be applied to SAS programming.

How Is Efficiency Defined?

The most basic definition of efficiency is to get more results from fewer resources. In reality the gain is seldom absolute because optimizing one resource usually results in increased consumption of another resource. For example, re-creating a SAS data set in each run avoids the cost of storing the data set but increases the I/O and CPU time for the program.

This book defines *efficiency* as obtaining more results from fewer computer or human resources. Therefore, it deals with two components of efficiency: computer efficiency and human efficiency.

Computer Efficiency

When you submit a SAS program, the computer system must

- [ ] load the required software into memory

- [ ] compile the program

- [ ] find the data on which the program will execute

- [ ] perform the calculations or other operations requested in the program

- [ ] report the result of the operations for your inspection.

All of these tasks require time and space. Time and space for a computer program are composed of CPU time, I/O time, and memory.

CPU time is the time the *CPU*, or Central Processing Unit, spends performing the calculations or other operations you assign. In this book, techniques that decrease CPU time are marked with the icon shown here.

I/O time is the time the computer spends on two tasks, input and output. Input refers to moving data from storage areas such as disks or tapes into memory for work, and output refers to moving the results out of memory to storage or to a display device such as a terminal or a printer. In this book, techniques that decrease I/O time are marked with the icon shown.

Memory is the size of the work area that the CPU must devote to the operations in a program. In this book, techniques that decrease memory are marked with the icon shown at left.

Another important resource is data storage.

Data storage refers to how much space on disk or tape your data use. In this book, techniques that decrease data storage are marked with the icon shown.

This book defines computer efficiency in terms of these resources rather than in terms of elapsed time (also known as throughput), that is, how quickly you get results. *Elapsed time* is composed of computer time and waiting time. Computer time refers to the items discussed earlier in this section, such as I/O time and CPU time. Waiting time includes waiting for programs to run, waiting for CPU resources, waiting for results to print, and so forth. Waiting time is largely a measure of the resources available to you: the number of jobs being processed at the same time as yours, how much reading and writing they do, the type of computer hardware at your site, and many other factors. This book concentrates on showing you how to make the best use of the resources you can control.

Human Efficiency

The final type of efficiency this book discusses is human efficiency.

Human efficiency is how much programming time is required, both to develop and to maintain a program. Programming time in this sense includes all the time a programmer spends working on a program: designing, coding, looking up information, testing, debugging, documenting, and so forth. In this book, techniques that decrease the programming time are marked with the icon shown.

Tradeoffs

Improving a program's efficiency requires making decisions about *tradeoffs*, that is, which resources to optimize and which to use less efficiently. The SAS System illustrates a tradeoff between computer and human resources. Most of the default actions of the SAS language and procedures reduce human workloads by letting the computer do more of the work, thereby increasing the consumption of computer resources.

When you decide to improve a program's efficiency, you choose to invest programming time in overriding default settings in order to save either machine resources or long-term human resources in maintaining the program. In addition, a few programming techniques do improve performance in almost all areas. Such blue-ribbon techniques are described here.

Certain tips decrease the use of several resources and have few, if any, disadvantages. You can use them to improve the efficiency of almost any application. Those tips are marked with the icon shown at left.

If your site charges for computer resources, the best set of tradeoffs is the one that produces the lowest charge for your programs.

How Do You Decide What Is Important for Efficiency?

When you consider the number of elements that affect the efficiency of a program and the necessity of making tradeoffs, you must decide which factors are the most important for improving the use of resources in your own work. To make the decision, you must take three steps:

□   Understand efficiency at your site.

□   Know how your program will be used.

□   Know the type of data your program will process.

The rest of this section discusses these factors.

Understand Efficiency at Your Site

The possibilities for improving efficiency in programs vary widely. They depend on the type of programming you do and also, more importantly, on the computing environment at your site. Environmental factors affecting the efficiency of SAS programs include the following:

hardware
> The hardware environment at your site has several aspects. The most obvious factor is the capacity of your hardware, including the amount of main memory you have available for processing. (*Main memory* is the area that a CPU can access rapidly and from which the CPU executes instructions and operates on data.) Other factors include the number of terminals, printers, and other devices attached to the CPU; the communications hardware in use; and the number of users that share that hardware.

host operating system
> Factors affecting your host operating system include how it has been customized. In particular, the resource allocation and scheduling algorithm and the method of input and output may vary at different sites.

SAS environment
> The SAS environment at your site encompasses which SAS software products have been installed; how they have been installed; which related applications software products have been installed, such as various sorting utilities; and which methods are available to run SAS programs at your site. Only the SAS Software Representative can change these aspects of your SAS environment. Another important aspect is the default settings of various SAS system options at your site. You can override the settings of some SAS system options.

In most cases one or two resources will be the most limited or most expensive for your programs. For example, if you work with extremely large data sets that must reside on disk, storage may be your biggest concern. If your processing budget is limited, CPU time may be most important. You can usually decrease the amount of critical resources used if you are willing to give up some efficiency of resources that are less critical at your site.

Know How Your Program Will Be Used

Developing any program requires time. Developing an efficient program can be partly automatic as you acquire efficient programming habits, but it usually requires some additional time and thought. Therefore, the first question to ask yourself is whether the amount of resources saved is worth the time and effort spent to achieve the savings.

To answer the question, first consider size. Size in this case can refer to either the number of statements in the program or the size of the files being processed. For a small program or file, the absolute difference between an inefficient and an efficient program is not great. As the programs or files get larger, the potential for savings increases. Therefore, you should devote more effort to making large programs efficient.

Also consider the number of times the program will run. The difference in the resources used by an inefficient and an efficient program run once or a few times is relatively small, whereas the cumulative difference for a program run frequently (for example, every night for a year) is large.

Both of these factors operate in the same direction. The importance of efficiency increases both as the size of the program or file increases and as the number of executions of the program increases. Therefore, you should invest the least effort in optimizing small programs that will run few times and the greatest effort in optimizing large programs that will run many times.

Know the Type of Data Your Program Will Process

The effectiveness of any efficiency tip depends heavily on the data with which you use it. For example, a tip to speed up the processing of missing values is most effective when data contain many missing values. When you know the characteristics of your data, you can select tips that take advantage of those characteristics.

How Do These Principles Apply?

This section shows two case studies that apply the general principles of efficiency to specific cases. The case studies show how investing some extra time in programming can pay off with a large increase in the computer efficiency of the program. Part 2, "Tips for Efficient SAS Programming," contains detailed suggestions for techniques you can use in your programs.

Case 1: Speeding Up a Repeated Task

A medical researcher is studying the effect of a new drug on patients. The researcher keeps the patients' medical histories in a large SAS data set with many observations. Because few people leave or enter the study, the file undergoes few changes. The researcher regularly extracts small subsets of patients by age, previous treatments, and other factors and merges those with the test data to perform analyses. The researcher would like to speed up the process of selecting subsets.

Because the budget for the study is limited and the most expensive part of the analysis is creating the subsets, the researcher decides to save CPU time and I/O time by creating indexes for the large SAS data set and using the WHERE statement rather than a subsetting IF statement to select the subsets. The researcher trades a slight increase in storage space for storing the indexes for a large decrease in the time required to create the subsets and, therefore, a lower overall cost.

Case 2: Preventing Unnecessary Processing

An insurance company has a production job that updates several master files with the day's transactions every night. Because the job involves large files and runs frequently, the programmer in charge of the process wants to make it run as efficiently as possible. The current program has two problems:

□ Although the job edits the transactions before performing the updates, serious errors occasionally corrupt the master files. In that case the programmer must restore the master files from backup, allow users to correct the transactions, and update the master files with the corrected transactions the following day.

□ All updates run every night, although sometimes there are no transactions in a particular category. In that case updating a particular master file is unnecessary.

To make the job more efficient, the programmer can write an editing program that sends observations with serious errors to a separate SAS data set. The update program can then check the number of observations in the errors data set before beginning the updates. If the errors data set contains any observations, the update program can stop processing at that point. The next day users can correct the transactions and the programmer can run the update program without having to restore files from backup. By adding the additional DATA step to the program, the programmer can save both computer and human resources.

In addition, the programmer can have the editing program write each category of transactions to a separate SAS data set. The update program can determine whether a particular transaction data set has any observations before beginning to update a particular master file. When there are no transactions, the program does not begin the DATA step that updates that master file. The programmer can decrease the computer time spent in compiling and executing the large updates by adding small DATA steps to determine whether a SAS data set contains any transactions.

Chapter 2 How Do You Know if Your Program Is Efficient?

Introduction

Since efficiency is a relative term, you can only measure how efficient your programs are by comparing the resources they consume with some standard. You should evaluate different techniques when you want to improve the efficiency of large jobs or jobs you run frequently. To make an evaluation, you need to collect information about resource usage as you test different versions of the same program.

This chapter discusses the SAS tools available for collecting performance and storage statistics, provides some guidelines for testing programming techniques, and discusses evaluating the results of those tests.

Collect Performance Statistics

Performance statistics include measures of performance showing the number of resources your SAS program uses when it compiles and executes. The most commonly used measures of performance include the items in the following table.

Resource	How Measured
CPU time	seconds or fractions of a second
I/O operations	number of times a program requests a read or write operation between main memory and a storage device
memory	the maximum number of bytes of actual main memory the compilation and execution of the program used

You can use the following two SAS system options to ask for measures of CPU time, I/O operations, and memory under any operating system:

□ STIMER

□ FULLSTIMER.

The way you specify these system options varies by operating system. On some operating systems, you must also specify other SAS system options to get performance statistics in the SAS log. The performance statistics you get in the SAS log vary by operating system. See the SAS companion for your operating system for a full description of the contents of the SAS log and for the SAS system options you must specify to get performance statistics in the SAS log at your site.

Once you have performance statistics in your SAS log, you can then begin testing different versions of your programs to identify more efficient techniques.

▶ *Caution* . . . . *Turn off the performance statistics after you finish testing since reporting them consumes resources.* ▲

In addition to the SAS system options you can set to display measures of performance in the SAS log, the SQL procedure has an STIMER option that shows performance measures on a query-by-query basis. Refer to *SAS Guide to the SQL Procedure: Usage and Reference, Version 6, First Edition* for more information.

For a complete description of all SAS system options available in base SAS software, as well as a description of the SAS log, refer to *SAS Language: Reference, Version 6, First Edition*.

Collect SAS Data Set Information

SAS data set information related to data storage includes anything that measures how much space a SAS data set occupies. You can ask for generic and operating-system-specific information regarding the contents of a SAS data set. This information relates both to the space a data set occupies and to the number of I/O operations you generate during processing when you read the data set.

Report Data Set Statistics with Utility Procedures

The CONTENTS procedure, or the CONTENTS statement in the DATASETS procedure, produces the most comprehensive data set statistics you can get from base SAS software, including number of variables, variable types, observation length, number of indexes, and compression status. For a list of items contained in the output of the CONTENTS procedure, see "Tip 9.7: Document your SAS data sets."

You can use the CONTENTS procedure to produce a report or an output data set. Producing an output data set with the CONTENTS procedure allows you to track statistics about data sets automatically using the SAS System to accumulate the results directly.

Interpret the Output

Although the output of the CONTENTS procedure provides a wealth of information regarding your SAS data sets, you must interpret the results. Consider other factors to see if your data are being stored in the least amount of space. Ask yourself the following questions:

□ Can you use character variables instead of numeric variables?

□ Can you reduce the length of some numeric variables and preserve the precision you need?

□ Can you get your character data into shorter variables?

□ Will compressing your data set save space?

There are other data storage statistics, including data set page size, that you can apply in ways that vary depending on your operating system. Data set *page* size tells you how much data the SAS System moves in one input/output operation, and you can use it to evaluate storage requirements as well as CPU and I/O statistics. See the SAS companion for your operating system for more details about interpreting and using these statistics to improve the efficiency of your data storage.

Verify Programming Techniques

Although applying the tips provided in this book should improve the efficiency of your SAS programs, exactly how much improvement you notice depends on your application, your system configuration, and how you apply the techniques. Given the performance and data storage information you collect, you can determine the degree to which any given technique improves your processing efficiency.

The following sections define benchmarking, discuss factors that affect the resources your programs use, and show ways to reduce the influence of those factors when testing programs and analyzing results.

Benchmark Your Code

Measuring the difference in resources used by different programming techniques is a special form of testing called *benchmarking*. The principle behind benchmarking programming techniques is to test identical pieces of code, varying only the specific technique whose effect you want to measure.

You can only benchmark code on a particular machine under known conditions. Benchmarks valid on one hardware configuration may vary dramatically when you perform them on other equipment.

Most of the tips in this book provide two pieces of example code: one using a technique that works and another using a technique that works better. Benchmark your own code using a similar technique, and vary only the particular statement, option, or order you want to test with a benchmark. Varying other factors complicates the issue since many changes may cancel each other out. One change may produce a neutral effect and another a positive effect, but if you benchmark them both at once, you don't know which change gained the efficiency you measured.

Account for All Factors

There are many factors that affect the amount of improvement that you can expect from applying any of the efficiency tips in this book. These factors also influence the results you get from benchmarking code. They include the following:

□ the size and type of your computer system

□ your operating system

□ the method you use to communicate with your computer system

□ the number of records in files you process or the number of observations in SAS data sets you process

□ the resource allocation and scheduling algorithm your computer system uses

□ the way that the SAS System is installed at your site

□ the method you use to run SAS software at your site

□ other applications you have installed at your site.

Since these factors can interact with each other in many different ways, any efficiency gains you record from applying the tips in this book can vary even between different processing runs on the same computer system. Try the techniques in this book to see if they save you enough resources to justify using them in particular situations.

Use Consistent Methods

The following sections give some guidelines for obtaining consistent results from program benchmarking at your site.

Control Your Code

Separate your steps clearly at each *step boundary*, or end of a complete DATA or PROC step, with RUN statements so that the statistics in the SAS log immediately follow the steps you are testing. For procedures that support RUN groups, include QUIT statements, especially if you benchmark in interactive line mode or with the SAS Display Manager System.

Eliminate as much irrelevant or peripheral code from your samples as possible. This action focuses the results of the benchmark on the technique you test rather than on extra DO loops, CALL routines, macro code, or other elements that do not vary between your code samples.

Control Your Environment

Run your benchmarks under conditions that are as nearly alike as you can make them. Some things to consider include the following:

□ other users on your system

□ other resource-intensive processing on your system

□ file differences

□ invocation differences

□ method of operation (batch versus interactive line mode, for example).

Use Multiple Runs

Avoid the temptation of basing conclusions on a single processing run. Run each program several times; run each iteration from the invocation of the SAS System through the end of the session. Gather the statistics you need from the SAS logs and calculate averages.

Use Single-Purpose Runs

Dedicate SAS sessions you use to benchmark a particular technique to just that purpose. If you run benchmarks during a session that is already underway, the processing that already occurred during the session will affect the results of the benchmark. Thus, you should start a new SAS session for each program you want to test.

Remember the Effect of Overhead

Any time you invoke the SAS System, use particular DATA step features, or invoke a particular procedure within a SAS session, there is a load *overhead*, or additional resources associated with getting the code from disk into main memory. To accurately benchmark a technique, you must use code samples that are identical, except for the technique you are testing, in order to account for load overhead.

Explore Other Statistics

Although the SAS System provides tools for obtaining performance and data storage statistics, there are other tools available that vary by operating system. Consult the documentation provided by your hardware and software vendors to see if those tools can meet your benchmarking needs.

Make Prudent Conclusions

If techniques that you test do not show dramatic differences in resource usage, don't automatically discard them. The resource savings they achieve may be evident under circumstances different than those under which you tested. For example, they may be more efficient with larger data sets, with more numeric variables than character variables, or with steps using more file operations. However, if these other factors will not materialize in your application, the technique may not save any resources in your case.

Similarly, if a technique saves resources in your tests, remember that it was more efficient given the circumstances under which it was tested. Varying other factors may alter any efficiency advantage a technique may have.

Part 2

Tips for Efficient SAS® Programming

Part 2 presents programming techniques arranged by the general SAS programming principle they illustrate. The first chapter in this section introduces seven general principles for improving the efficiency of SAS programs. Other chapters contain a series of tips that illustrate one of these general principles.

Chapter **3** How Can Your SAS® Programs Be More Efficient?

Introduction

The SAS System is powerful, flexible, and easy to program. Its design maximizes the computing power available to you while minimizing your workload in programming. At the same time, this design introduces the possibility that SAS programs can produce the correct results while consuming far more resources than necessary in the process.

This chapter introduces seven general principles of efficient programming that you can use in saving computer and human resources. The last section of this chapter illustrates how to read a typical programming tip.

What Are the Principles of Efficient Programming?

The first step in improving your SAS programs is to recognize that there may be areas in which you can increase their performance. To help you identify those areas, this book presents seven general principles to follow in writing more efficient SAS programs. Each chapter in Part 2, "Tips for Efficient SAS Programming," is organized around one of the following principles:

□ **Read and write data selectively.** The fewer variables and observations you process, the fewer I/O operations the SAS System must do.

□ **Execute only the statements you need, in the order you need them.** Executing the minimum number of statements, in the most efficient order, minimizes the CPU time the SAS System uses.

□ **Take advantage of SAS procedures.** Different procedures are designed for different tasks, and using one designed for your task is more efficient than using one that just happens to work. In addition, when a DATA and a PROC step can produce the same result, you need to know which method is more efficient.

□ **Know SAS System defaults.** In order to bypass SAS System defaults for particular purposes, you must understand default values and actions, and know how to change them.

□ **Control sorting.** Rearranging data is one of the most common operations in the SAS System, and knowing how and when to sort makes a big difference in the CPU time consumed in a program.

□   **Know your data and test your programs.** Being familiar with your data and programs is a major step in general programming efficiency. If you know your data, you know the kind of data errors that may be present and the kind of analyses that are possible. By verifying that your programs work correctly, you minimize the time spent later in correcting errors and in rerunning programs that produce incorrect output.

□   **Code clearly.** Coding programs clearly and documenting them with comments minimizes the maintenance time for you or another programmer later.

These principles are valid principles for general SAS programming. However, remember that the effectiveness of any tip depends heavily on the computing environment at your site and the processing requirements of your program. You should test all tips at your site before incorporating them into your regular programming.

What Is an Efficiency Tip?

Each chapter is composed of modules showing a specific programming technique, or tip, that implements the general principle. The following list describes the parts of a programming tip. Each part is indicated by a corresponding boldface number in the sample tip shown in Figure 3.1.

1   The tip number identifies the tip for reference, and the short title briefly states the tip.

2   The icon or icons indicate the resources saved. The first icon listed indicates the largest saving, and the others follow in order.

3   The text of the tip gives more information and, if needed, a description of the programming statements needed to implement it.

4   "Using This Tip" first describes the feature of the SAS System that makes the tip work and then describes any circumstances in which the tip is particularly helpful. Finally, this section describes any tradeoffs of using the tip.

5   The module usually ends with either a comparison of acceptable and more efficient code or another type of illustration. Red code may be used to emphasize SAS statements that illustrate the tip.

Figure 3.1
A Sample Tip

1

Tip 4.7: **Create all data subsets at one time.**

2 | CPU | I/O |

3 Minimize the number of times you read large SAS data sets or external files by producing all of the subsets you require for further processing in one DATA step. Test for conditions using IF/THEN statements and write observations to multiple data sets using OUTPUT statements.

4 **Using This Tip**

You can conditionally output to any number of data sets at a single pass of a large data set or external file if you structure your conditions correctly and specify several data sets in the DATA statement. You read each observation in the large data set or each record in the external file just once, but output the observations to as many data sets as you need.

5

Acceptable	More Efficient

```
data a;
   set master;
   if score1;
run;

data b;
   set master;
   if score1 and age lt 20;
run;

data c;
   set master;
   if score1 and age ge 20;
run;
```

```
data a b c;
   set master;
   if score1 then
      do;
         output a;
         if age lt 20
            then output b;
         else output c;
      end;
run;
```

Chapter **4** Read and Write Data Selectively

Understanding the Principle

Reading and writing data (called input/output or I/O) is the largest single component of elapsed time in most programs, including SAS programs. Decreasing the number of I/O operations is the single most effective way to improve the execution speed of your programs. Most methods for reducing the number of I/O operations fall into one of the following broad categories:

☐ reading or writing based on a condition, rather than reading or writing all the time

☐ simplifying the information you read or write, or reducing it in size

☐ substituting execution time transformations for large reads or writes

☐ eliminating redundant read or write steps with better logic

☐ making use of SAS features and tools.

The tips in this chapter give several concrete ways to read and write data selectively and more efficiently.

Tip 4.1:   Read only the fields you need.

When you read an external file, use pointer controls, informats, or column specifications in the INPUT statement to read only those fields you actually need.

Using This Tip

When you read records from an external file in the DATA step, you spend CPU time when you read a field. If you use list input without pointer controls, you must read every field in an external file into a corresponding SAS variable. When you use list input with pointer controls, formatted input, or column input, you exclude unneeded variables from your data set and spend less CPU time reading fields you don't need.

Acceptable

```
data oct90;
   infile file-specification;
   input jday rcode $ sales exps
         salhrs exphrs invin
         invout;
run;
```

More Efficient

```
data oct90;
   infile file-specification;
   input jday 1 sales 5-13
         invout 42-50;
run;
```

Tip 4.2: Read selection fields first.

Determine if you can eliminate records based on the contents of one or two fields. Read those fields first with an INPUT statement, using an @ line-hold specifier to hold the record, and test for a qualifying value. If the value meets your criteria, read the rest of the record with another INPUT statement. Otherwise, delete the record from the program data vector.

Using This Tip

When you create a SAS data set from an external file, you can just read the entire file and use the SAS System to create subsets of the data set later as you need them. However, by screening the contents of one or two fields, you eliminate the need to subset the data later, and only keep relevant data in storage. Using this tip always saves resources with no tradeoffs or disadvantages.

Acceptable

```
data year90.sales;
   infile file-specification;
   input month sales3
         sales18 . . .;
   more SAS statements
run;

data aug90.sales;
   set year90.sales;
   if month=8;
run;

proc means data=aug90.sales;
   more SAS statements
run;
```

More Efficient

```
data aug90.sales;
   infile file-specification;
   input month 1 @;
   if month=8 then
      do;
         input sales3 3-9 . . .;
         more SAS statements
      end;
   else delete;
   more SAS statements
run;

proc means data=aug90.sales;
   more SAS statements
run;
```

Tip 4.3: Store data in SAS data sets.

When you need to use the SAS System repeatedly to analyze or manipulate any particular group of data, create a permanent SAS data set instead of reading the raw data each time to create a temporary SAS data set.

Using This Tip

With your data already in a permanent SAS data set, you can use any SAS procedure, function, or routine on the data without any further conversion. Starting with SAS data sets saves CPU time and the extra I/O operations associated with reading a raw data file. SAS data sets are also self-documenting and enable you to include additional documentation with data set labels, variable labels, variable formats and informats, descriptive variable names, and so on.

Acceptable

```
data in;
   infile file-specification;
   input . . .;
   more SAS statements
run;

proc means data=in;
   more SAS statements
run;
```

More Efficient

```
libname mylib
         'SAS-data-library';

proc means data=mylib.in;
   more SAS statements
run;
```

Tip 4.4: Keep summaries of large SAS data sets.

For frequent or repeated reporting or analysis, use the SUMMARY procedure to create permanent summary SAS data sets from large SAS data sets or SAS data sets on tape. Determine the lowest level of summarization you need, then build small summary SAS data sets to contain that information.

Using This Tip

When you have a large SAS data set and you need to run multiple reports on specific subsets of the data, read the large data set once. Create permanent summary data sets containing just the information you need for each series of reports. For data sets on tape, creating permanent summary data sets on disk makes applications run faster because you eliminate the time it takes to mount and read the tape for each run.

Acceptable

```
proc summary
    data=big.dataset nway;
   class x y;
   var a;
   output out=summary;
run;

data report;
   set summary;
   put x y a=;
   more SAS statements
run;

proc summary
    data=big.dataset nway;
   class x y;
   var b;
   output out=summary;
run;

proc chart data=summary;
   vbar b;
run;
```

More Efficient

```
proc summary
    data=big.dataset nway;
   class x y;
   var a b;
   output out=small.summary;
run;

data report;
   set small.summary;
   put x y a=;
   more SAS statements
run;

proc chart data=small.summary;
   vbar b;
run;
```

Tip 4.5: Store only the variables you need.

Store only the variables you need by using DROP or KEEP statements, DROP= or KEEP= data set options, or constants instead of variables.

Using This Tip

There are variables you need only during DATA step execution. Some examples are

□ index variables from DO loops

□ variables holding intermediate values in calculations

□ variables holding values for testing conditions.

Eliminating these variables from the output data set saves the disk space you need to store them.

<table>
<tr><td>

Acceptable

```
data short.rpt;
   infile file-specification;
   retain datx '17JUL90'd;
   input charges 1-8 code $ 9
         date mmddyy8.;
   if upcase(code)='A' and
      date lt datx then
      do i=1 to 5;
      more SAS statements
run;
```

</td><td>

More Efficient

```
data short.rpt;
   infile file-specification;
   drop code i;
   input charges 1-8 code $ 9
         date mmddyy8.;
   if upcase(code)='A' and
      date lt '17JUL90'd then
      do i=1 to 5;
      more SAS statements
run;
```

</td></tr>
</table>

Tip 4.6:   Process only the variables you need.

Use the DROP= or KEEP= data set option with the input SAS data set in the SET, MERGE, or UPDATE statement to keep unneeded variables out of the logical program data vector.

Using This Tip

When you refer to an input data set using the SET, MERGE, or UPDATE statement, select the variables you need using the DROP= or KEEP= data set option rather than the DROP or KEEP statement. When you use data set options to select the variables you want, you eliminate all of the other variables from the logical program data vector as well as from the output data sets you produce.

You can use the DROP or KEEP statement in combination with the DROP= or KEEP= data set option to

□ eliminate variables created in the current DATA step from the output data set

□ eliminate variables that you need in the program data vector during the current DATA step from the output data set.

The more variables there are in the input SAS data set, the more resources this tip saves.

Acceptable

```
data stocks.gainers;
   set stocks.oct;
   keep numshrs company;
   if net gt 0 then
      do;
         tot=net*numshrs;
         put company numshrs;
         output;
      end;
run;
```

More Efficient

```
data stocks.gainers;
   set stocks.oct
         (keep=numshrs company net);
   drop net;
   if net gt 0 then
      do;
         tot=net*numshrs;
         put company numshrs;
         output;
      end;
   run;
```

Tip 4.7: Create all data subsets at one time.

Minimize the number of times you read large SAS data sets or external files by producing all of the subsets you require for further processing in one DATA step. Test for conditions using IF/THEN statements and write observations to multiple data sets using OUTPUT statements.

Using This Tip

You can conditionally output to any number of data sets at a single pass of a large data set or external file if you structure your conditions correctly and specify several data sets in the DATA statement. You read each observation in the large data set or each record in the external file just once, but output the observations to as many data sets as you need.

Acceptable	More Efficient

```
data a;
   set master;
   if score1;
run;

data b;
   set master;
   if score1 and age lt 20;
run;

data c;
   set master;
   if score1 and age ge 20;
run;
```

```
data a b c;
   set master;
   if score1 then
      do;
         output a;
         if age lt 20
            then output b;
         else output c;
      end;
run;
```

Tip 4.8:

Use informats for data transformations.

Define and store your own informats for common data transformations using the FORMAT procedure. Specify your informats in the INPUT statement when reading data into SAS data sets rather than transforming data with programming logic.

Using This Tip

When you read external files, you may need to translate words or numeric value ranges into codes or make other translations. Translations like these are called *data transformations*. You can transform the data using programming logic. However, by letting an informat read the data, you save the CPU time associated with successive IF/THEN logic and the programmer time associated with coding and maintaining the more complex code. You also eliminate extra variables in the program data vector even if you drop them in the data set.

Example Data

```
THREE     ONE       ONE
THREE     THREE     ONE
TWO       TWO       TWO
ONE       THREE     TWO
more data lines
```

Acceptable

```
data in;
   infile file-specification;
   input x $10. y $10. z $10.;
   drop x y z;
   if upcase(left(x))='ONE'
      then xnum=1;
   else if upcase(left(x))='TWO'
      then xnum=2;
   more SAS statements
run;
```

More Efficient

```
libname library
        library-specification;

   /* Run this step once to */
   /* create the informat.  */
proc format library=library;
   invalue $onetwo (just upcase)
      'ONE'=1 'TWO'=2 'THREE'=3;
run;

data in;
   infile file-specification;
   input xnum $onetwo10.
         ynum $onetwo10.
         znum $onetwo10.;
   more SAS statements
run;
```

Tip 4.9: Shorten data using formats and informats.

Use the FORMAT procedure to create formats and informats for displaying and storing long character values. Use FORMAT or ATTRIB statements to associate your formats and informats with variables during processing.

Using This Tip

Define your own formats and informats when you can use a short code to represent long character data. Use informats to transform long character values in raw data into codes with shorter values in your SAS data sets. Use formats that convert coded values to longer values when you need them.

For example, you can use your informats to convert survey responses or inventory part names into codes. You store and process coded values in your data set using your informats, but can convert to the longer values in procedure output, or when you use the PUT function in a DATA step, using your formats. Although storing formats and informats requires some disk space, you save space if your data sets are large and long values are repeated over thousands of observations. By using this tip, you trade an increase in CPU time for a decrease in storage requirements.

Acceptable

```
data inv.partrec;
   infile file-specification;
   input slsman $ qty;
   select(slsman);
      when('FISHER')
        part='GALVANOMETER';
      when('SUAREZ')
        part='COMPRESSION FILTER';
      more SAS statements
run;
```

More Efficient

```
proc format library=library;
   invalue $slname
           'FISHER'='F'
           'SUAREZ'='S';
   value $slname
           'F'='FISHER'
           'S'='SUAREZ';
   invalue $pname
           'F'='BG' 'S'='PCF';
   value $pname
           'BG'='GALVANOMETER'
           'PCF'='COMPRESSION FILTER';
run;

data inv.partrec;
   infile file-specification;
   input slsman : $slname. qty;
   part=input(slsman,$pname.);
   attrib slsman format=$slname.
          part format=$pname.;
   more SAS statements
run;
```

Tip 4.10:   **Edit external files directly.**

Use the PUT statement with pointers to modify external files directly instead of converting raw data to SAS variables first and then editing variables with functions.

Using This Tip

When you need to modify fields in the records of an external file (for example, an input file), you can usually save CPU time by avoiding conversion of the data to SAS variables. This tip works best if you need to simply replace individual characters in a file at specific positions. You can construct conditions so that the code affects only specific records. This tip works especially well when you modify records that are longer than the maximum 200-byte length of a SAS character variable.

▶ *Caution* . . . . *Exercise caution when reading and writing to external files directly to avoid overwriting data by mistake.* ▲

The code in the following example places the character X in positions 10, 300, and 500 of a 600-byte record for all records in a file.

Acceptable

```
data _null_;
   infile 'file-specification';
   input c1 $char200. c2 $char200.
      c3 $char200.;
   file longr;
   substr(c1,10,1)='X';
   substr(c2,100,1)='X';
   substr(c3,100,1)='X';
   put (c1-c3) ($char200.);
run;
```

More Efficient

```
data _null_;
   infile 'file-specification';
   input;
   file longr;
   put _infile_ @10 'X'
               @300 'X'
               @500 'X';
run;
```

Tip 4.11: Use one buffer for external file operations.

Use the SHAREBUFFERS option in the INFILE statement, and specify the same fileref in the INFILE and FILE statements to create only one buffer in main memory when you read from and write to the same external file.

Using This Tip

Specifying the SHAREBUFFERS option in the INFILE statement eliminates the need for separate input and output buffers. A *buffer* is a temporary storage area reserved for holding data after they are read or before they are written.

▶ *Caution . . . . Exercise caution when reading and writing to external files directly to avoid overwriting data by mistake.* ▲

The code in the following example places the character X in positions 10, 300, and 500 of a 600-byte record for all records in a file.

Acceptable

```
data _null_;
   infile longr;
   input;
   file longr;
   put _infile_ a10 'X'
               a300 'X'
               a500 'X';
run;
```

More Efficient

```
data _null_;
   infile longr sharebuffers;
   input;
   file longr;
   put a10 'X'
       a300 'X'
       a500 'X';
run;
```

Tip 4.12: Create indexes when appropriate.

Create indexes for SAS data sets with the DATASETS procedure or the SQL procedure when

□ the data set is relatively large

□ the values within the data set are not frequently updated

□ the data set is frequently subset by values of the indexed variable

□ the data within the data set are uniformly distributed

□ a typical data subset yields less than one-third of the observations.

Using This Tip

Indexed SAS data sets can provide significant performance improvements for large data sets. For relatively small data sets, sequential processing is often just as efficient as indexed processing.

When you have a data set that frequently changes, the overhead associated with rebuilding an index after each update can outweigh the processing advantages you gain from accessing the data through an index. In the following example, the more efficient step becomes less efficient if you must reindex the data set each time you need a data subset.

Acceptable

```
data stasu.dnoterec;
   set stasu.dnote3;
   if in_date gt '14APR90'd;
run;
```

More Efficient

```
   /* Run this step once to */
   /* create the index.    */
proc datasets library=stasu;
   modify dnote3;
   index create in_date;
run;

data stasu.dnoterec;
   set stasu.dnote3;
   where in_date gt '14APR90'd;
run;
```

Tip 4.13:   Use binary search, not null merge.

Use the POINT= option in the SET statement to locate a set of observations in a master data set that correspond to observations having a common variable in the transaction data set.

Using This Tip

When you have a large, sorted master data set and a small, unsorted transaction data set, use a binary search to locate observations with common BY variable values in the master. A common solution to this problem uses a MERGE statement with the IN= option and a BY statement in a DATA _NULL_ step. Using binary search, you skip a sort and only search half or less of the master file sequentially. Using the MERGE statement, you read both data sets record by record.

The larger the master data set you have, the better this tip works.

Acceptable

```
    /* Null Merge Technique */
proc sort data=trans;
   by id;
run;

data _null_;
   merge master(in=inm)
         trans(in=int)
         end=endf;
   by id;
   if int then do;
      if inm then do;
         put / 'FOUND: '
             @15 id=
             @30 name=;
         fnd+1;
      end;
      else do;
         put / 'NOT FOUND: '
             @15 id=;
         nfnd+1;
      end;
   end;
   if endf then do;
      put / 'TOTAL FOUND: '
          @40 fnd;
      put / 'TOTAL INVALID: '
          @40 nfnd /;
   end;
run;
```

More Efficient

```
    /* Binary Search Technique */
data _null_;
   set trans
       (rename=(id=trid))end=endf;
   low=1;
   flag=1;
   up=nobs;
   do while(flag);
      mid=ceil((up+low)/2);
      set master point=mid
                 nobs=nobs;
      if id<trid then low=mid+1;
      else if id>trid
              then up=mid-1;
      else do;
         put /'FOUND: '@15 trid=
                       @30 name=;
         fnd+1;
         flag=0;
      end;
      if low>up then do;
         put / 'NOT FOUND: '
             @15 trid=;
         nfnd+1;
         flag=0;
      end;
   end;
   if endf then
      do;
         put / 'TOTAL FOUND: '
             @40 fnd;
         put / 'TOTAL INVALID: '
             @40 nfnd /;
      end;
run;
```

Chapter 5 Execute Only the Statements You Need, in the Order You Need Them

Understanding the Principle

The number and complexity of statements executed in a particular step largely control the CPU time used in the step. By default, the SAS System executes every statement in the DATA step for each observation in the input source. Within a statement, it executes every operation in a given expression each time the statement executes. Tips in this chapter fall into two main categories:

☐ tips that reduce the number of statements executed

☐ tips that reduce the number of operations performed in a particular expression.

To determine when to use the tips in this chapter, ask yourself the following questions:

□ How many observations from the input source do I need in this DATA step? For how long? How many observations, if any, do I need to write to the output data set?

□ How many statements will be executed on a particular observation?

□ How many parts of an expression need to be executed for a particular statement?

The remainder of this chapter is a series of tips that illustrate specific ways to reduce the number of statements and operations executed and to execute them in the most efficient order.

Tip 5.1:   Assign a value to a constant only once.

Assign a constant in a RETAIN statement instead of in an assignment statement.

Using This Tip

The SAS System assigns values to variables in a RETAIN statement once, during DATA step compilation. *Compilation* is the automatic translation of SAS statements into executable code that usually occurs before your program runs. In contrast, the SAS System executes an assignment statement during each iteration of the DATA step. If you place a constant in an assignment statement, the SAS System assigns the same value to the variable many times.

Acceptable

```
data scores;
   infile file-specification;
   input test1-test3;

      /* Same values for X & Y */
      /* in every iteration    */
   x=5;
   y="&sysdate"d;
run;
```

More Efficient

```
data scores;

      /* Same values for X & Y */
      /* in every iteration    */
   retain x 5 y "&sysdate"d;
   infile file-specification;
   input test1-test3;
run;
```

Tip 5.2: Use constants in expressions.

Use constants or macro variable references that resolve to constants in expressions rather than numeric variables.

Using This Tip

During each execution of an expression, the SAS System checks to see whether the value of each variable in the expression is missing. It does not check constants. Therefore, you can save CPU time by using constants in expressions.

In previous releases of the SAS System, you could increase efficiency by assigning a constant used more than once to a variable and using the variable as needed. That technique is obsolete. In Release 6.06, repeating the constant is more efficient.

Acceptable	**More Efficient**

```
%let base=1000;
libname save 'SAS-data-library';

data big;
   retain b &base;
   set save.overnite;
   goal=productn+b;
   lowpt=min( b,avg);
   more SAS statements
run;
```

```
%let base=1000;
libname save 'SAS-data-library';

data big;
   set save.overnite;
   goal=productn+&base;
   lowpt=min( &base,avg);
   more SAS statements
run;
```

Tip 5.3: Condense constants in expressions.

When an expression involves two or more constants with one or more arithmetic operators or the logical operators AND or OR, write the expression so that the constants occur in sequence.

Using This Tip

During compilation, the DATA step evaluates parts of expressions involving only constants. During execution, the DATA step uses only the result of the evaluation. Therefore, evaluating expressions or parts of expressions at compilation eliminates that evaluation in each iteration of the DATA step during execution.

Acceptable

```
%let today1=520;
%let today2=650;
%let today3=800;
%let today4=149;
%let today5=720;

   /* Add 10 terms    */
   /* in each iteration */
data current;
   infile file-specification;
   input base1-base5;
   total=base1 + &today1
        +base2 + &today2
        +base3 + &today3
        +base4 + &today4
        +base5 + &today5;
run;
```

More Efficient

```
%let today1=520;
%let today2=650;
%let today3=800;
%let today4=149;
%let today5=720;

   /* Add 6 terms     */
   /* in each iteration */
data current;
   infile file-specification;
   input base1-base5;
   total=base1 + base2
        +base3 + base4
        +base5 + &today1
        +&today2 + &today3
        +&today4 + &today5;
run;
```

Tip 5.4:   **Use mutually exclusive conditions.**

When only one condition can be true for a given observation, write either a series of IF-THEN/ELSE statements or a SELECT group rather than a series of IF-THEN statements without ELSE statements.

Using This Tip

In IF-THEN/ELSE statements or SELECT groups, the SAS System stops checking statements when a condition is true for an observation. It then skips to the end of the series and resumes processing. However, in a sequence of IF-THEN statements without ELSE statements, the SAS System checks each condition for every observation. This tip is most useful when some categories contain many more observations than other categories and you can place the mutually exclusive conditions in order of descending probability. (See "Tip 5.5: Write conditions in order of descending probability.")

In the DATA step in Release 6.06, IF-THEN/ELSE statements and SELECT groups (with or without a SELECT expression) use equivalent CPU resources.

Acceptable

```
data planes;
   input number city $;
   if city='Rome' then flight=654;
   if city='Dakar' then
      flight=230;
   if city='Riga' then flight=165;
   cards;
6602 Rome
8304 Dakar
5990 Riga
4598 Rome
2300 Rome
1600 Dakar
9400 Rome
2812 Riga
3379 Rome
;
```

More Efficient

```
   /* choose one method: */
1 data planes;
   input number city $;

      /* IF-THEN/ELSE    */
   if city='Rome' then flight=654;
   else if city='Dakar' then
        flight=230;
   else if city='Riga' then
        flight=165;
   cards;
data lines
;


2 data planes;
   input number city $;

      /* SELECT without a  */
      /* SELECT expression */
   select;
      when(city='Rome')
          flight=654;
      when(city='Dakar')
          flight=230;
      when(city='Riga')
          flight=165;
   end;
   cards;
data lines
;


3 data planes;
   input number city $;

      /* SELECT with a SELECT */
      /* expression           */
   select (city);
      when('Rome')
          flight=654;
      when('Dakar')
          flight=230;
      when('Riga')
          flight=165;
   end;
   cards;
data lines
;
```

Tip 5.5: Write conditions in order of descending probability.

In a series of mutually exclusive IF-THEN/ELSE statements or in a SELECT group, place the most likely condition first. Continue with conditions of descending probability and place the least likely condition last.

Using This Tip

In a series of mutually exclusive conditions, the SAS System stops evaluating conditions after it finds a true one. Placing the most likely conditions at the top of the list causes the SAS System to execute the fewest statements, thus saving CPU time. A related example is shown in "Tip 5.4: Use mutually exclusive conditions."

Acceptable	More Efficient

```
libname in 'SAS-data-library';


data india1 italy1 iceland1;
   set in.worldpop;


      /* India is most likely */
   select(country);
      when('Iceland') output
         iceland1;
      when('Italy') output
         italy1;
      when('India') output
         india1;
      otherwise;
   end;
run;
```

```
libname in 'SAS-data-library';


data india1 italy1 iceland1;
   set in.worldpop;


      /* India is most likely */
   select(country);
      when('India') output
         india1;
      when('Italy') output
         italy1;
      when('Iceland') output
         iceland1;
      otherwise;
   end;
run;
```

Tip 5.6:   Put missing values last in expressions.

Write expressions so that the operations most likely to involve missing values are performed last.

Using This Tip

When a SAS expression contains several operations, a missing value propagates from the first operation in which it occurs through all subsequent operations in the expression. The SAS System records the column and line location of each use of the missing value and how many missing values occur at that location. (By default, the information appears in messages in the SAS log, but the SAS System maintains the information regardless of whether you print or suppress log notes.) The fewer operations that involve missing values, the less record-keeping the SAS System must do.

Acceptable

```
data test;
   infile file-specification;
   input t1-t5;

      /* t1 is often missing */
   score=t1+t2/2+t3/3+t4/4+t5/5;
run;
```

More Efficient

```
data temp;
   infile file-specification;
   input t1-t5;

      /* t1 is often missing */
   score=t2/2+t3/3+t4/4+t5/5+t1;
run;
```

Tip 5.7:   Check for missing values before using a variable in multiple statements.

If you use a variable that often has missing values in several assignment statements, check the variable for a missing value in an IF condition and place a DO group containing the assignment statements in the THEN clause. If the value of the variable is not missing, execute the DO group.

Using This Tip

Propagating missing values in expressions requires CPU time, as discussed in "Tip 5.6: Put missing values last in expressions." By checking for a missing value before performing the operations, you can reduce the number of missing values that are propagated. In particular, remember that the SAS System sets the value of a variable created in an assignment statement to missing at the beginning of the DATA step. If the value of the new variable is calculated from a missing value, the SAS System simply replaces one missing value for another. If you prevent the calculation by checking for missing values, the SAS System retains the missing value assigned at the beginning of the DATA step and your program uses less CPU time.

Acceptable

```
/* OFTMISS is often missing */
/* Missing value propagates */
/* and is assigned to COST, */
/* TAX, and PROFIT          */
data a;
    infile file-specification;
    input oftmiss wholsale sales;
    cost=wholsale+oftmiss;
    tax=oftmiss*.05;
    profit=sales-oftmiss;
run;
```

More Efficient

```
/* OFTMISS is often missing  */
/* When OFTMISS is missing,   */
/* use default missing values*/
/* for COST, TAX, & PROFIT-- */
/* bypass propagation         */
data a;
    infile file-specification;
    input oftmiss wholsale sales;
    if oftmiss ne . then
        do;
            cost=wholsale+oftmiss;
            tax=oftmiss*.05;
            profit=sales-oftmiss;
        end;
run;
```

Tip 5.8: **Put only statements affected by the loop in a loop.**

Put only statements that are affected by the change of the loop's index variable or expression inside iterative DO, DO WHILE, or DO UNTIL loops.

Using This Tip

When you have an iterative DO, DO UNTIL, or DO WHILE loop, check to be sure that all statements within it need to be executed during each iteration of the loop. A loop executes all the statements within it each time it iterates, and during each iteration, the value of an index variable or an expression changes. Statements that are not affected by that change do not need to be executed repeatedly; if they are, the SAS System simply replaces the same value each time.

Acceptable

```
data record;
   input store $ invntry avgsales
         year;
   if invntry<10000 then
      do;
         put 'Reorder now: '
             store= invntry=;
         delete;
      end;
   else
      do month=1 to 12
         while(invntry>10000);
         restock=invntry-avgsales;
         if year=. then year=1990;
      end;
   keep store month year;
   cards;
Raleigh 16000 500 1989
Cary 18000 1000 .
Wilson 9000 600 1990
;
```

More Efficient

```
data record;
   input store $ invntry avgsales
         year;
   if year=. then year=1990;
   if invntry<10000 then
      do;
         put 'Reorder now: '
             store= invntry=;
         delete;
      end;
   else
      do month=1 to 12
         while(invntry>10000);
         restock=invntry-avgsales;
      end;
   keep store month year;
   cards;
Raleigh 16000 500 1989
Cary 18000 1000 .
Wilson 9000 600 1990
;
```

Tip 5.9:   Assign many values in one statement.

Use the FORMAT procedure with a VALUE statement to create formats representing the values you want to assign. Then use the PUT function with the format in an assignment statement to assign the new values. If you plan to use the format repeatedly, store it in a permanent SAS data library.

Using This Tip

When you must change many values of a variable to other values, using user-created formats with the PUT function enables you to make the changes in a single assignment statement. You do not need to test multiple conditions to find which one is true before changing the value. In addition, if the possible values change, you can maintain the program by changing the PROC FORMAT step rather than changing every occurrence of the variable in the program.

Acceptable	More Efficient

```
libname in 'SAS-data-library';

data new;
   set in.finance;
   length money $ 7;
   if country='US' then
      money='Dollar';
   else if country='Mexico' then
      money='Peso';
   else if country='Japan' then
      money='Yen';
   else if country='Israel' then
      money='Shekel';
   else if country='Spain' then
      money='Peseta';
   else if country='Greece' then
      money='Drachma';
   else if country='India' then
      money='Rupee';
   else if country='Iceland' then
      money='Koruna';
run;
```

```
libname in 'SAS-data-library';

proc format;
   value $money
         'US'='Dollar'
         'Mexico'='Peso'
         'Japan'='Yen'
         'Israel'='Shekel'
         'Spain'='Peseta'
         'Greece'='Drachma'
         'India'='Rupee'
         'Iceland'='Koruna';
run;

data new;
   set in.finance;
   money=put(country,$money.);
run;
```

Tip 5.10:   Shorten expressions with functions.

Use functions, when available, rather than coding your own expressions.

Using This Tip

Functions provided with SAS software use precompiled expressions. Therefore, a DATA step containing a function needs to compile only the name of the function and its arguments. If you write an expression to do the same thing, the SAS System must compile all operators and operands in the expression. In addition, a function may execute faster than your expression. Finally, SAS functions are tested as part of the development process of SAS software, whereas you must spend time programming and debugging expressions you create.

Acceptable

```
/* mean of non-missing costs */
/* note programming for       */
/* missing values             */
data getmeans;
   array c{10} cost1-cost10;
   input c{*};
   tot=0;
   totcost=0;
   do i=1 to 10;
      if c{i} ne . then
         do;
            tot+1;
            totcost+c{i};
         end;
   end;
   if tot>0 then
      meancost=totcost/tot;
   else meancost=.;
   cards;
1 2 3 4 5 6 7 8 9 10
10 . 30 . 50 . 70 . 90 .
;
```

More Efficient

```
/* mean of non-missing costs */
/* missing values are handled*/
/* automatically             */
data getmeans;
   input cost1-cost10;
   meancost=mean(of cost1-cost10);
   cards;
1 2 3 4 5 6 7 8 9 10
10 . 30 . 50 . 70 . 90 .
;
```

Tip 5.11: Edit character values with functions.

Use functions to edit character values instead of breaking them apart and concatenating the pieces. Useful functions include the COMPRESS, INDEX, LEFT, REVERSE, RIGHT, SUBSTR, TRANSLATE, TRIM, UPCASE, and VERIFY functions.

Using This Tip

Character functions can remove or replace unwanted characters or change the alignment of the value. Using character functions enables you to create fewer variables and avoid executing IF-THEN statements.

Acceptable

```
   /* get last name */
data lastname;
   length last $ 25;
   input name $ 1-25;
   n=1;
   do until(scan(name,n,' ')=' ');
      n+1;
   end;
   last=scan(name,n-1,' ');
   cards;
Christopher A. Jones
Ann Siu
Ernst Otto Jacob Grosz
Julia Lincoln Romero
;
```

More Efficient

```
   /* get last name */
data lastname;
   length last $ 25;
   input name $ 1-25;
   last=reverse(scan
        (reverse(name),1,' '));
   cards;
Christopher A. Jones
Ann Siu
Ernst Otto Jacob Grosz
Julia Lincoln Romero
;
```

Tip 5.12: Use the IN operator rather than logical OR operators.

When comparing a value to a series of constants, use the IN operator rather than a series of OR operators.

Using This Tip

When the SAS System evaluates an expression containing the IN operator, it stops the evaluation as soon as a comparison makes the expression true. When the SAS System evaluates an expression containing multiple OR operators, it evaluates the entire expression even if one true comparison has already made the expression true. In addition, the IN operator is simpler to write and to maintain than a series of OR operators.

Acceptable

```
data address;
   infile file-specification;
   input street $ number;
   if street='Maple' or
      street='Elm' or
      street='Willow' or
      street='Birch' or
      street='Dogwood' or
      street='Ash' then
      city='Raleigh';
   else city='Cary';
run;
```

More Efficient

```
data address;
   infile file-specification;
   input street $ number;
   if street in
      ('Maple','Elm','Willow',
       'Birch','Dogwood','Ash')
      then city='Raleigh';
   else city='Cary';
run;
```

Tip 5.13: Use a series of conditions.

When a statement contains a large number of conditions, use a series of IF-THEN clauses rather than a compound expression with AND.

Using This Tip

When a DATA step contains a series of conditions in IF-THEN clauses, the DATA step stops evaluating the series as soon as one clause is false. However, when the conditions are joined by AND, the DATA step evaluates all of the conditions even if one is false.*

This tip is most useful for statements with lengthy conditions. Note that a series of IF-THEN clauses is more difficult for a programmer to read than conditions joined by AND. Also, ELSE statements are difficult to position properly after a series of IF-THEN clauses.

Acceptable

```
data jobs;
    infile file-specification;
    input status1-status8 jobnum;
    if status1=3 and
        status2=7 and
        status3=9 and
        status4=1 and
        status5=44 and
        status6=6 and
        status7=4 and
        status8=1 then output;
    drop status1-status8;
run;
```

More Efficient

```
data jobs;
    infile file-specification;
    input status1-status8 jobnum;
    if status1=3 then
      if status2=7 then
        if status3=9 then
          if status4=1 then
            if status5=44 then
              if status6=6 then
                if status7=4 then
                  if status8=1 then
        output;
    drop status1-status8;
run;
```

---

* The SQL procedure stops evaluating both case expressions and a series of conditions joined by AND when one condition is false.

Tip 5.14:   Check for undesirable conditions and stop processing.

In a production job, add code to detect conditions that would make output unusable. Then write an explanatory message with a PUT statement and end the program with the ABORT statement.

Using This Tip

A production job often contains certain points at which definite actions must take place to make finishing the program worthwhile. By stopping the job when those conditions are not met, you can avoid useless results. In addition, you may save CPU time or I/O operations.

For example, updating a master file and producing reports on the new master file are not necessary when there are no transactions. You can save resources by stopping the job before the update. Messages distinguish this type of ending from endings caused by system errors.

Acceptable

```
libname in 'SAS-data-library-1';
libname new 'SAS-data-library-2';

data in.master;
   update in.master new.trans;
   by transid;
run;

more DATA and PROC steps
```

More Efficient

```
libname in 'SAS-data-library-1';
libname new 'SAS-data-library-2';

data _null_;
   if number=0 then
      do;
         file print;
         put '**************' /
             '* no changes *' /
             '* job ends   *' /
             '**************';
         abort return;
      end;
   stop;
   set new.trans nobs=number;
run;

data in.master;
   update in.master new.trans;
   by transid;
run;

more DATA and PROC steps
```

Tip 5.15: Write the loop with the fewest iterations outermost.

In nested iterative DO loops, write the loop with the fewest changes in the value of the index variable outermost and the loop with the most changes innermost.

Using This Tip

When you nest two or more iterative DO loops, the total number of iterations is the same regardless of which loop is outermost. The formula for calculating the total number of iterations in a system of nested loops is

iterations of inner loop x   iterations of outer loop

However, the number of times the index variables must be incremented to achieve that number of iterations varies according to the relative positions of the loops. An inner loop cycles through all values of its index variable each time the outer loop takes on a single value. In addition, the outer loop cycles through all values of its index variable once. The formula for calculating the number of changes in index variables in a system of nested loops is

(iterations of inner loop x iterations of outer loop) + iterations of outer loop

The more iterations the outer loop contains, the more times the SAS System must increment the value of its index variable.

Acceptable	More Efficient

```
       /* 20,000 iterations */
       /* 30,000 increments */
    data loops;
       do a=1 to 10000;
         do b=1 to 2;
             more SAS statements
         end;
       end;
    run;
```

```
       /* 20,000 iterations */
       /* 20,002 increments */
    data loops;
       do b=1 to 2;
         do a=1 to 10000;
             more SAS statements
         end;
       end;
    run;
```

Tip 5.16:

Use temporary arrays rather than creating and dropping variables.

Assign constants to elements in _TEMPORARY_ arrays rather than to variables you drop with a DROP statement.

Using This Tip

You can create an array either of variables or of temporary data elements (often referred to as a _TEMPORARY_ array). _TEMPORARY_ arrays are more efficiently processed than arrays made of variables for two reasons:

□ _TEMPORARY_ array elements require about 30 bytes per element less overhead than do DATA step variables.

□ _TEMPORARY_ array elements are always contiguous in memory, whereas variables may or may not be. The SAS System can access contiguous memory directly, but it must access an array of variables through a series of steps, as shown:

TEMPORARY array
   array reference → data

array of variables
   array reference → table of pointers → variable name → data

If you create an array only to hold a group of constants that you do not need in the output data set, creating a _TEMPORARY_ array is more efficient.

Acceptable

```
data scores;
   infile file-specification;
   input test1-test6;
   array test{6} test1-test6;

   /* Needed for calculation */
   array cutoff{6} s1-s6
      (100 90 80 70 60 50);
   do i=1 to 6;
     if test{i}>=cutoff{i} then
        output;
   end;
   drop s1-s6;
run;
```

More Efficient

```
data scores;
   infile file-specification;
   input test1-test6;
   array test{6} test1-test6;

   /* Needed for calculation */
   array cutoff{6} _temporary_
      (100 90 80 70 60 50);
   do i=1 to 6;
     if test{i}>=cutoff{i} then
        output;
   end;
run;
```

Tip 5.17:   **Use macros for repeated code.**

Write repeated code in a macro and invoke the macro where it is needed in the DATA step instead of linking to groups of statements with LINK and RETURN statements.

Using This Tip

When the same code occurs repeatedly in a DATA step, you can save programming time by not repeating the code. Two ways to execute repeated code are to use LINK and RETURN statements or to use the macro facility. The macro facility is more efficient because when you invoke a macro at a particular point in a DATA step, the macro processor produces SAS statements at that location in the step. In contrast, using LINK and RETURN statements requires branching to another location in the DATA step, executing a group of statements, and returning to the original location. You can also tailor the code produced by each invocation of a macro through parameters. With subroutines, you must write a separate subroutine for each variation in the code.

Tradeoffs are that the SAS System uses space in the WORK library for a macro catalog and that making the macro facility available causes the SAS System to use slightly more memory.

Acceptable

```
data cities;
   infile file-specification;
   input oldpop city $ newpop;
   if oldpop>=800000 then link
      more;
   if city='London' then link
      more;
   if newpop>1000000 then
      do;
         link more;
         link most;
      end;
   return;

   more:
   SAS statements
   return;

   most:
   additional SAS statements
   return;
run;
```

More Efficient

```
%macro more(add);
   SAS statements
   %if &add ne %then
      %do;
         additional SAS statements
      %end;
%mend more;

data cities;
   infile file-specification;
   input oldpop city $ newpop;
   if oldpop>=800000 then

      /*DO group allows for    */
      /*multiple SAS statements*/
      /*generated by the macro */
      do;
         %more()
      end;
   if city='London' then
      do;
         %more()
      end;
   if newpop>1000000 then
      do;
         %more(yes)
      end;
run;
```

Tip 5.18: Create macro variables only when needed.

Use a CALL SYMPUT statement only when needed so that it doesn't execute during each iteration of the DATA step.

Using This Tip

By default, a CALL SYMPUT statement assigns a value to a macro variable during each iteration of the DATA step. If you want to assign the value from only one observation (for example, the last observation) to the macro variable, execute the CALL SYMPUT statement only during that iteration of the DATA step.

Acceptable	More Efficient

```
data _null_;
   set huge;
   income+(bonds*.06);
   call symput('inc',income);
run;
```

```
data _null_;
   set huge end=last;
   income+(bonds*.06);
   if last then
      call symput('inc',income);
run;
```

Tip 5.19:   **Put a variable into only one array.**

List a particular variable in only one ARRAY statement.

Using This Tip

Placing the same variable in more than one array increases CPU time because the SAS System must locate the variable in the additional arrays indirectly.

Acceptable	More Efficient

```
data children;
   infile file-specification;
   input c1-c10;

      /* half in group 1 */
   array group1{5} c1-c5;

      /* all in group 2  */
   array group2{10} c1-c10;
   do i=1 to 5;
      group1{i}=group1{i}+10;
   end;
   do j=1 to 10;
      group2{j}=group2{j}+10;
   end;
   more SAS statements
run;
```

```
data children;
   infile file-specification;
   input c1-c10;

      /* half in group 1 */
   array group1{5} c1-c5;

      /* half in group 2  */
      /* use array bounds */
      /* for convenience  */
   array group2{6:10} c6-c10;
   do i=1 to 5;
      group1{i}=group1{i}+20;
   end;

      /* change bounds of DO */
      /* loop to match array */
   do j=6 to 10;
      group2{j}=group2{j}+10;
   end;
   more SAS statements
run;
```

Tip 5.20:   Make array variables all retained or all unretained.

Use a RETAIN statement or assign initial values in the ARRAY statement to make all elements in an array either retained or unretained.

Using This Tip

The SAS System allocates separate blocks of memory for retained and unretained variables. Therefore, if some array elements are retained and some are not, all elements of the array cannot be contiguous in memory and the SAS System must access the elements indirectly, through pointers.

Acceptable	More Efficient

```
    /* two retained, two not */
data income;
   retain tax 600 fee 45;
   array pay{4} tax fee rent
        lease;
   infile file-specification;
   input rent lease;
   do i=1 to 4;
      if pay{i}=. then pay{i}=0;
      pay{i}=pay{i}*1.05;
   end;
run;
```

```
    /* all retained */
data income;
   retain tax 600 fee 45 rent
      lease;
   array pay{4} tax fee rent
        lease;
   infile file-specification;
   input rent lease;
   do i=1 to 4;
      if pay{i}=. then pay{i}=0;
      pay{i}=pay{i}*1.05;
   end;
run;
```

Tip 5.21:   Set the lower bound of arrays to 0.

When an array is frequently accessed, specify the array bounds as $\{0:n-1\}$ rather than the default $\{n\}$ or $\{1:n\}$, where n is the number of elements in the dimension. Make the range of the index variable in any iterative DO loops that process the array 0 to $n-1$.

Using This Tip

Internally, the SAS System begins counting the elements in an array with 0, not 1. Thus, all internal array bounds are 0 to $n-1$ rather than the 1 to n that you see by default. When you use the default array bounds, the SAS System subtracts 1 from the subscript of each array reference each time the reference is executed in order to locate the correct member of the array. Specifying bounds that match the internal bounds eliminates the calculation.

Note that one array reference executed once constitutes one access of the array. Therefore, the number of accesses in one DATA step is the number of accesses in one iteration times the number of iterations (usually the number of observations).

Programs in which array bounds start at 0 are slightly harder for a programmer to read than programs using the default bounds. In addition, the ranges of index variables of iterative DO loops that process the array must match the specified bounds.

Acceptable

```
    /* up to 20,000 subtractions */
data new;
    set old(keep=t1-t50 base
        obs=100);
    array test(50) t1-t50;
    do i=1 to 50;
        if test(i)=. then test(i)=0;
        if test(i)<base then
            test(i)=base;
    end;
run;
```

More Efficient

```
    /* no subtractions */
data new;
    set old(keep=t1-t50 base
        obs=100);
    array test(0:49) t1-t50;
    do i=0 to 49;
        if test(i)=. then test(i)=0;
        if test(i)<base then
            test(i)=base;
    end;
run;
```

Tip 5.22: Use multidimensional explicit arrays.

Use multidimensional explicit arrays to process variables as a table rather than an array whose elements are implicit arrays (that is, a multidimensional implicit array).

Using This Tip

Multidimensional explicit arrays are much more efficient than multidimensional implicit arrays because the SAS System can address elements in a multidimensional explicit array more directly. The following shows how the SAS System locates data values referenced in multidimensional explicit and implicit arrays:

□ Multidimensional explicit arrays can be composed of either variables or temporary data elements. The SAS System retrieves data through a multidimensional explicit array as follows:

array of variables
array reference → table of pointers → variable name → data

TEMPORARY array
array reference → data

□ The SAS System retrieves data through a multidimensional implicit array as shown:

multidimensional implicit arrays (array of arrays)
array reference → pointer to outer array → pointer to inner arrays → variable names → data

Multidimensional explicit arrays, whether made of variables or of temporary data elements, require fewer steps to locate data than multidimensional implicit arrays require.

Acceptable

```
data new;
   infile file-specification;
   input t1q1-t1q10 t2q1-t2q10
         t3q1-t3q10;
   array test1(j) t1q1-t1q10;
   array test2(j) t2q1-t2q10;
   array test3(j) t3q1-t3q10;
   array alltest(i) test1-test3;
   do i=1 to 3;
      do j=1 to 10;
         if alltest=. then
            alltest=0;
      end;
   end;
run;
```

More Efficient

```
data new;
   infile file-specification;
   input t1q1-t1q10 t2q1-t2q10
         t3q1-t3q10;
   array alltest(3,10) t1q1-t1q10
                       t2q1-t2q10
                       t3q1-t3q10;
   do i=1 to 3;
      do j=1 to 10;
         if alltest(i,j)=. then
            alltest(i,j)=0;
      end;
   end;
run;
```

Tip 5.23:   **Use the Stored Program Facility.**

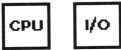

Use the Stored Program Facility to store and execute compiled code for complex DATA steps that execute repeatedly.

Using This Tip

When a DATA step contains many statements or complex statements and is executed repeatedly, using the Stored Program Facility eliminates the CPU time needed for compilation. Storing compiled DATA steps is possible because compiling and executing DATA steps are separate phases of processing.

When using this tip, you must save a copy of the source code for reference. Also note that in Release 6.06, a stored compiled DATA step does not contain its date and time of creation. Therefore, if you change the source code, you must manually keep track of whether the compiled code is up-to-date. Both of these steps increase the programming time needed to maintain the program.

Acceptable

```
libname in 'SAS-data-library-1';

data longjob1;
   set in.region1;
   more SAS statements
run;

data longjob2;
   set in.region2;
   same SAS statements
run;
```

More Efficient

```
libname in 'SAS-data-library-1';
libname save 'SAS-data-library-2';

   /* compile and store */
   /* SAVE SOURCE CODE! */
data longjob1;
   set in.region1;
   more SAS statements
run pgm=save.storejob;

libname in 'SAS-data-library-1';
libname save 'SAS-data-library-2';

   /* execute stored program */
data pgm=save.storejob;
   redirect input in.region1=
      in.region1;
   redirect output longjob1=
      longjob1;
run;

   /* execute again */
data pgm=save.storejob;
   redirect input in.region1=
      in.region2;
   redirect output longjob1=
      longjob2;
run;
```

Chapter 6 Take Advantage of SAS® Procedures

Understanding the Principle

SAS procedures are built-in programs that perform a wide variety of specialized tasks. Most procedures either perform several different tasks within a general category or perform the same task several different ways. Different procedures can perform similar tasks, but each procedure performs more efficiently in specific circumstances. A few procedures in base SAS software, like the SQL procedure, have their own unique language, syntax, and, in some cases, data storage mechanisms.

Most of the techniques for taking advantage of SAS procedures to improve your programming efficiency fall into one of the following broad categories:

☐ using procedures instead of code you develop yourself

☐ using the right procedure for the right task

☐ using a procedure to perform as many different tasks as possible once you have invoked it

☐ using the output of procedures for other tasks

☐ knowing the full capabilities of any given procedure

☐ knowing all of the SAS tools you can use for any given task.

The tips in this chapter give a variety of resource-saving features of SAS procedures.

Tip 6.1:   Let procedures do the work.

Take advantage of the variety and depth of SAS procedures to process jobs you may only need to perform a limited number of times rather than developing DATA step or macro code yourself.

Using This Tip

When you want information from data quickly, you can usually find a SAS procedure to provide the information. Even though you may prefer customized results, it may not be worth your time to develop your own DATA step or macro code if you only need the results a limited number of times.

Acceptable

```
data new(drop=count score i);
   array scores{4} col1-col4;
   retain count 1 scores;
   set exams;
   by id;
   if first.id then
      do i=1 to 4;
         scores{i}=.;
      end;
   scores{count}=score;
   count+1;
   if last.id then
      do;
         count=1;
         output;
      end;
run;

proc print data=new;
   title 'DATA step transpose';
run;
```

More Efficient

```
proc transpose data=exams out=new;
   by id;
run;

proc print data=new;
   title 'Procedure transpose';
run;
```

Tip 6.2: Take advantage of output SAS data sets.

Use the output SAS data sets from the CORR, FREQ, and MEANS procedures as input to other procedures to avoid reading a master data set again. Use options with the MEANS, UNIVARIATE, and COMPARE procedures to control the contents of the SAS data set each produces. Use the output data set from the CONTENTS procedure to pass information about data set and variable characteristics across step boundaries. Use the output data set from the COMPARE procedure to determine the degree of difference between data sets using SAS programs.

Using This Tip

SAS procedures produce output SAS data sets for one of the following reasons:

- □ to provide a utility to produce an output data set, as in the APPEND, STANDARD, and TRANSPOSE procedures

- □ to provide a summary, *data reduction*, or special form of a data set that you can use as input to other procedures, as in PROC MEANS, PROC FREQ, and PROC CORR

- □ to provide an alternative to replacing the current data set with processed data, as in the SORT procedure

- □ to provide an accessible repository of information you can use to read in other steps to make processing decisions, as in PROC COMPARE and PROC CONTENTS

- □ to provide the raw data behind the typical output, enabling you to use or summarize the data as you want rather than relying on just the standard procedure output, as in PROC UNIVARIATE and PROC COMPARE.

In many cases, you can use the output data set that a procedure produces as an alternative to more expensive processing. In some cases, the output data set is the only way to communicate information between program steps.

Tip 6.2   *continued*

The following table lists several base SAS procedures that produce output
SAS data sets, and provides a short description of what each contains.

Procedure	Output SAS Data Set Contents
APPEND	concatenated SAS data sets, or just a copy of the input data set if you specify no base data set
COMPARE	observations for each pair of matching observations in the comparison SAS data sets with variables describing the differences; observations for each pair of matching numeric variables in the comparison data sets with variables describing the differences
CORR	special type SAS data sets with different correlation statistics that other SAS/STAT procedures recognize as input
FREQ	one observation for each combination of the variable values in the last table request, with extra variables that give the frequency and cell percentage for the combination of variable values (other procedures can use this output SAS data set as input)
MEANS/SUMMARY	with CLASS statement, observations that summarize the combinations of class variables; other descriptive summary variables (other procedures can use this output SAS data set as input)
RANK	ranks or rank scores of the variables you specify in the input SAS data set
SORT	all the observations and variables in the input SAS data set, but in the order you specified; when you specify an output data set with PROC SORT, the input data set remains intact
STANDARD	variables containing the standardized values for the variables you specify in the input data set
TRANSPOSE	observations correspond to variables in the input SAS data set, with variables copied from the input data set, variables created by transposing observations, and identification variables created by the procedure
UNIVARIATE	one observation for each unique set of values of the variables in the BY statement, or a single observation if there is no BY statement; variables contain values specified when you invoke the procedure

For a more thorough discussion of all procedure output SAS data sets, see the
SAS Procedures Guide, Version 6, Third Edition.

Tip 6.3:   Use procedures to examine your data.

Use the CHART procedure with the HBAR statement to see the big picture. Use the PLOT procedure to spot relationships among data variables. Use the MEANS procedure to see variable ranges and limits. Use the COMPARE procedure to test for differences between SAS data sets.

Using This Tip

Each SAS procedure has an important role to play in exploratory data analysis. Knowing the characteristics of your data can help you process it more efficiently, since you can analyze those elements that show the most promise and immediately eliminate apparently unproductive research directions.

Examples of procedures you should use to your advantage in exploring your data include the following:

CHART procedure

HBAR charts of character values highlight bad or unexpected values, for example, illegal codes. HBAR charts of numeric values help identify extreme values or outliers and their approximate magnitude. With the GROUP= option, HBAR charts can show tendencies within groups of related variables.

PLOT procedure

By plotting the values of variables against one another, you can spot key relationships easily. If the values of variables vary over a wide range, compress the plot by making one or both scales logarithmic.

MEANS procedure

The N option gives the number of observations; the NMISS option gives the number of observations with missing values; the MIN and MAX options give the smallest and largest values of variables; the SUM, MEAN, and STD options give the sum, mean, and standard deviation of all numeric variables.

COMPARE procedure

You can compare entire SAS data sets or just the values of particular variables. Use PROC COMPARE to see if one processing technique provides exactly the same output as another.

Tip 6.4: **Copy indexed SAS data sets with procedures.**

Use the COPY statement in the DATASETS procedure rather than a SET statement in a DATA step to make a simple copy of an indexed SAS data set.

Using This Tip

When you use the SET statement in a DATA step to make a copy of a SAS data set, the copy does not contain the data set label or any of the indexes associated with the original data set. The COPY statement in the DATASETS procedure preserves data set labels and indexes, while using the SET statement in a DATA step does not.

The COPY statement in PROC DATASETS also provides the flexibility to copy several SAS data sets at one time, including entire SAS data libraries. Additionally, you can copy library members that are not data sets, or entire categories of members, with the MEMTYPE= option.

Copying indexed SAS data sets with the SET statement is a two-step process since you must re-create the index with PROC DATASETS, as shown in the following acceptable example. The more efficient example demonstrates copying with PROC DATASETS in only one step, since you copy all of the SAS data sets and their indexes at the same time.

Acceptable

```
   /*Copy the data set. */
data jacknew.v2;
   set jack.v2;
run;

   /* Recreate the index. */
proc datasets library=jacknew;
   modify prods;
   index create
         ix1=(numvar1
              numvar2
              numvar3);
run;
```

More Efficient

```
   /* Copy the data set and */
   /* all of the indexes.   */
proc datasets;
   copy out=jacknew in=jack;
   select v2;
run;
```

Tip 6.5:   **Store formats with the SAS data sets that use them.**

When you define your own formats or informats with the FORMAT procedure, store the catalog that contains them in the same permanent library as the SAS data sets that use them.

Using This Tip

When you move permanent SAS data sets containing encoded or shortened data to other computers, you must make sure that the user-defined formats and informats you need to transform the data are transferred with them. Remembering the location of the catalog containing the user-defined formats and informats you need to process specific SAS data sets is easier if both the catalog and the data sets are in the same library.

Acceptable

```
libname permfmts
        'SAS-data-library-1';
libname movefmts
        'SAS-data-library-2';
libname permdata
        'SAS-data-library-3';
libname movedata
        'SAS-data-library-4';

proc datasets;
   copy out=movefmts in=permfmts
        memtype=catalog;
   copy out=movedata in=permdata
        memtype=data;
run;
```

More Efficient

```
libname permdata
        'SAS-data-library-1';
libname movedata
        'SAS-data-library-2';

proc datasets;
   copy out=movedata in=permdata
        memtype=(catalog data);
run;
```

Tip 6.6:   **Use WHERE conditions in procedures.**

Use a WHERE statement or a WHERE= data set option to run any procedure on a subset of a SAS data set. Use the WHERE statement to dynamically change the WHERE condition within RUN groups. Use the WHERE statement's special operators to increase the flexibility of your subsetting.

Using This Tip

When you want to run a procedure with only the observations in a data set that meet specific conditions, use the WHERE statement to select those observations rather than creating a subset data set and then running the procedure.

With the SAME AND operator, you can add conditions to an existing WHERE clause in interactive procedures that support RUN groups. By using a null WHERE statement, you can clear a WHERE condition and specify another condition within a RUN group, eliminating the need for multiple data subsets.

The WHERE statement supports a number of special operators, such as CONTAINS, IS NULL, LIKE, and sounds-like (=*), in addition to standard arithmetic and logical operators. These operators enable you to specify wildcard or pattern conditions that you cannot express any other way. A *wildcard* is a character used to represent possible characters at a particular position in a word in order to generalize the word.

Acceptable

```
    /* Create data subset. */
data temp;
   set time90.total;
   if month lt 4;
run;

    /* Run procedure. */
proc plot data=temp;
   plot var1*var2;
run;

    /* Create another subset. */
data temp;
   set time90.total;
   if month eq 7;
run;

    /* Run procedure again. */
proc plot data=temp;
   plot var2*var3;
run;
```

More Efficient

```
    /* Run procedure one time. */
proc plot data=time90.total;
   where month lt 4;
   plot var1*var2;
run;

    /* Use RUN groups for */
    /* other subsets.     */
   where month eq 7;
   plot var2*var3;
run;
quit;
```

Tip 6.7:

Use the SQL procedure to simplify your code.

Use the processing features of the SQL procedure to develop programs that contain fewer lines of code than traditional SAS programs.

Using This Tip

PROC SQL offers a simpler coding solution when you need to do the following:

□ use combinations of more than two SAS data sets

□ match on variables that are not exactly the same

□ calculate using intermediate results, as in percentage distributions.

Acceptable

```
data temp;
   set orders;
   cno=substr(cno,2);
   pno=substr(pno,2);
proc sort data=temp;
   by cno;
proc sort data=cust;
   by no;
data temp;
   merge temp(in=t)
      cust(in=c
      rename=(no=cno));
   by cno;
   if t and c;
proc sort data=temp;
   by pno;
proc sort data=parts;
   by no;
data temp;
   merge temp(in=t)
      parts(in=p
      rename=(no=pno));
   by pno;
   if t and p;
proc print;
run;
```

More Efficient

```
proc sql;
   select o.cno, c.name, o.pno,
          p.desc, o.qty
   from orders o, parts p, cust c
   where substr(cno,2)=c.no
         and substr(pno,2)=p.no;
run;
```

Tip 6.8: **Take advantage of column operations in the SQL procedure.**

Use the SQL procedure to perform operations among variables of different SAS data sets.

Using This Tip

The SQL procedure enables you to operate on specific variables of existing data sets without having to specify the DROP= or KEEP= SAS data set option to prevent reading the entire observation into the program data vector. You can match observations by the values of any variable without presorting any of the SAS data sets. When you use a WHERE clause to retrieve data, you can also limit the number of observations you process.

Using a WHERE clause in the SQL procedure works faster if there is an index on the variables you specify in the WHERE clause.

Acceptable	**More Efficient**

```
proc sort data=lib.puts;
   by stock;
run;

proc sort data=lib.calls;
   by stock;
run;

data lib.new;
   merge lib.puts lib.calls;
      by stock;
   keep tcost stock;
   tcost=(pprice*pnum)+
         (cprice*cnum);
run;

proc print data=lib.new;
run;
```

```
proc sql;
   select puts.stock, calls.stock,
          (pprice*pnum)+
          (cprice*cnum)
             as tcost
   from sql.puts, sql.calls
   where puts.stock=calls.stock;
```

Tip 6.9: ## Get cross-tabulations with output SAS data sets.

Use the CLASS statement and the OUTPUT statement with the MEANS procedure or the SUMMARY procedure to get summary cross-tabulations without reading your SAS data set more than once. Specify variables in the CLASS statement in the reverse order you want to see cross-tabulation combinations in the output data set. Use the NWAY option to limit the groupings in the output data set to just summaries of variable combinations.

Using This Tip

When you want to look at summary cross-tabulations for data sets with many classification variables, the output data set you get from using the CLASS and OUTPUT statements in PROC SUMMARY or PROC MEANS provides you the equivalent of multiple cross-tabulations with only one pass at your data.

Acceptable

```
proc sort data=listener;
   by sex music age;
run;

proc freq data=listener;
   by sex music age;
   tables sex;
   tables music;
   tables age;
   tables age*music;
   tables sex*age;
   tables sex*music*age;
run;
```

More Efficient

```
proc means data=listener;
   class sex music age;
   output out=sumdata;
run;

proc print data=sumdata;
run;
```

Chapter **7** Know SAS® System Defaults

Understanding the Principle

The design of the SAS System gives you maximum computing power with a minimum of coding. At the same time, you need to understand its default actions so you can override them when that suits your current needs.

The remainder of this chapter is a series of tips that illustrate specific situations in which overriding default actions of the SAS System can save you CPU time, storage space, and other resources.

Tip 7.1:   **Reduce the storage space for variables.**

Use the LENGTH statement to reduce the storage space for variables in SAS data sets.

Using This Tip

The SAS System uses default lengths for variables in SAS data sets unless you specify a different length. For character variables and for numeric variables containing integers, you can save significant storage space by specifying the length.

When using this technique, you must be careful to avoid truncating significant bytes. The problem is most common with numeric variables, but character variables can also be affected. See the SAS companion for your host operating system for the number of bytes required to store various integers accurately.

▶ *Caution* . . . . *Do not shorten numeric variables containing fractions.*
Storing fractions in fewer than 8 bytes may cause a significant loss of precision. See the section "Details of Numeric Precision" in Chapter 3, "Components of the SAS Language," in *SAS Language: Reference, Version 6, First Edition* for complete information. ▲

Using shortened numeric variables slightly increases the CPU time for a step. Because the SAS System uses 8 bytes for all numeric variables in the program data vector, the SAS System must expand shortened numeric variables when it moves them into the program data vector.

Acceptable	**More Efficient**
| ```
data report;
   input local catalog;
   sales=local+catalog;
   cards;
100 400
200 350
;
``` | ```
data report;
   length sales local catalog 4;
   input local catalog;
   sales=local+catalog;
   cards;
100 400
200 350
;
``` |

Tip 7.2:   Shorten SAS date values.

Use a LENGTH statement to store SAS date values in 4 bytes (5 bytes under the PRIMOS operating system).

Using This Tip

All SAS date values from approximately A.D. 1700 through A.D. 2200 are represented by five digits. Five-digit integers can be stored in 4 bytes on all operating systems under which the SAS System runs (except PRIMOS, which requires 5 bytes). Therefore, you should store all common SAS date values in variables of length 4 (5 under PRIMOS).

Expanding the shortened variables to 8 bytes in the program data vector requires additional CPU time. Also, if you transfer programs between another host operating system and PRIMOS, you should use a length of 5 to avoid truncation under PRIMOS.

Acceptable

```
data years;
   infile file-specification;
   input schedule;
   yr95='01jan95'd;
run;
```

More Efficient

```
data years;
   infile file-specification;
   input schedule;
      /* 4 or 5 bytes */
   length yr95 4;
   yr95='01jan95'd;
run;
```

Tip 7.3: Know the rules for creating null data sets.

Specify the reserved name _NULL_ in the DATA statement to have the SAS System go through a DATA step without creating a SAS data set.

Using This Tip

At times you need to perform DATA step processing without creating a SAS data set. For example, you may need to write reports with PUT statements or examine the data set's directory with the NOBS= option in the SET statement to determine the number of observations, which you then pass to another step.

In such cases, begin the DATA step with the DATA _NULL_ statement, not the DATA statement with no argument. The DATA _NULL_ statement causes the SAS System to process a DATA step without writing any observations to a SAS data set. In contrast, the DATA statement with no argument (no data set name or _NULL_) creates a SAS data set with the default name DATA1, DATA2, and so on.

Acceptable

```
data;
   set plants;
   file print;
   put a5 fertilzr a12 yield;
   more SAS statements
run;
```

More Efficient

```
data _null_;
   set plants;
   file print;
   put a5 fertilzr a12 yield;
   more SAS statements
run;
```

Tip 7.4: Avoid default type conversions.

Use the PUT function to perform numeric-to-character conversions, and use the INPUT function to perform character-to-numeric conversions.

Using This Tip

When you use a character variable containing digits in an arithmetic operation, the SAS System produces a temporary numeric value to use in the operation. When you use a numeric variable in a character operation (for example, producing a substring of digits in the value), the SAS System produces a temporary character value. These actions are examples of *default type conversions*.

In both cases, the SAS System must determine that the conversion is needed and determine the format to be used in creating the converted value. Performing the conversion explicitly rather than allowing the SAS System to perform it automatically removes those two steps.

Acceptable

```
/* change office addresses */
/* (OFFICE is character-- */
/* 101, 102, and so on)   */
data new;
   set old;
   office=office+100;
run;
```

More Efficient

```
/* change office addresses */
/* (OFFICE is character-- */
/* 101, 102, and so on)   */
data new;
   set old;
   office=put(input
             (office,4.)+100,4.);
run;
```

Tip 7.5: Use character rather than numeric variables.

If you do not plan to perform arithmetic operations on variables containing digits, store the variables as character rather than numeric.

Using This Tip

By default, numeric variables occupy 8 bytes of storage in a SAS data set, whereas character variables can occupy as few as 1 byte. In addition, all numeric variables occupy 8 bytes in the program data vector, whereas character variables can occupy as few as 1 byte. Storing up to 8 digits in character variables saves disk space.

This storage method requires extra processing if you do need to perform an arithmetic operation. You must either allow an automatic conversion to take place, or you must convert the value to a numeric one as shown in "Tip 7.4: Avoid default type conversions." Both of those actions increase the CPU time for the step, but the automatic method requires much more CPU time than the explicit method.

Acceptable	**More Efficient**
| ```
data inventry;
   input item start finish;
   sales=start-finish;
   cards;
1101 500 133
1102 650 105
1103 400  98
more data lines
;
``` | ```
data inventry;
   length item $ 4;
   input item $ start finish;
   sales=start-finish;
   cards;
1101 500 133
1102 650 105
1103 400  98
more data lines
;
``` |

Tip 7.6: Create separate variables instead of repeating type conversions.

To store the same information as both numeric and character, use two variables. Either read the input field twice, once as numeric and once as character, or perform one conversion with the PUT or INPUT function and save the resulting variable.

Using This Tip

Keeping two variables in a SAS data set saves CPU time that would otherwise be spent in making numeric-to-character and character-to-numeric conversions.

This method requires more storage space because you must store two variables instead of one. However, unless disk space is critical, one additional variable causes few problems. If you only need one form of the variable in a given step, you can drop the other.

Acceptable	More Efficient
| ```
data stock;
   length code $ 9;
   input price 1-3 item $ 5-8
        store $ 10-11;
   code=item||left(price)||store;
   ptax=price*1.06;
   keep ptax code;
   cards;
119 suit ak
220 coat pd
;
``` | ```
data stock;
   length code $ 9;
   input price 1-3 pchar $ 1-3
        item $ 5-8 store $ 10-11;
   code=item||pchar||store;
   ptax=price*1.06;
   keep ptax code;
   cards;
119 suit ak
220 coat pd
;
``` |

Tip 7.7: **Use SAS configuration and autoexec files.**

Put options that you use regularly into a SAS configuration file, and put statements that you use regularly into a SAS autoexec file.

Using This Tip

Even the best efficiency tips work only if you use them. Putting efficient default settings for system options into a configuration file ensures that you use them in each program and also saves you the time of entering them in each program. Similarly, you can store frequently used statements such as LIBNAME and FILENAME statements in an autoexec file to avoid entering them each time.

Note that storing your autoexec and configuration files requires some disk space.

The appearance and method of creating configuration and autoexec files vary according to your host operating system. See the SAS documentation for your host operating system for more information.

Tip 7.8:   **Compress large SAS data sets.**

Use the COMPRESS= data set option or system option when creating large SAS data sets if storage is a problem.

Using This Tip

Compressing a SAS data set means storing each observation in a shortened form. When compressing a data set, the SAS System ignores boundaries between variables within an observation and replaces repeated occurrences of a character with a two- or three-byte string. Compressed data sets require less storage space, and I/O operations are faster because a data set page can hold more compressed than uncompressed observations.

Compression is especially helpful when many contiguous variables have the same value, such as a series of 0s or blanks. If space is extremely limited, consider grouping variables that are likely to have the same value to maximize the effect of compression. Compression has more effect on short, wide data sets (those with few observations and many variables or several long character variables) than it does on long, narrow data sets (those with many observations and few variables) because the SAS System begins a new compression at the beginning of each observation.

You cannot use most forms of access by observation number with compressed SAS data sets, such as the POINT= option in the SET statement and commands used in various SAS/FSP procedures. The FIRSTOBS= and OBS= data set options and system options work with compressed data sets but more slowly than with uncompressed data sets.

In addition, the SAS System uses CPU time in preparing compressed observations for processing. Once compressed, you must re-create the data set with a DATA step or the DATASETS or COPY procedure in order to uncompress it.

Acceptable	**More Efficient**
`libname save 'SAS-data-library';`	`libname save 'SAS-data-library';`

`data save.huge;`
`  more SAS statements`
`run;`
	`data save.huge(compress=yes);`
	`  more SAS statements`
	`run;`

Tip 7.9:   Eliminate the macro facility if your programs do not need it.

If a program does not use any macro variables, macro windows, macros, or the DATA step interfaces SYMGET and SYMPUT, specify the NOMACRO system option at SAS invocation or in a configuration file.

Using This Tip

Removing the macro facility when a program does not require it reduces the compilation time of DATA and PROC steps because the SAS System does not keep the macro processor ready to check ampersands and percent signs that precede SAS names.

    If you modify the program later to use the macro facility, you must remember to specify the MACRO system option to make the macro facility available.

Tip 7.10: Store numbers as 1-byte character values.

Use the numbers 1 through 127 on ASCII systems or 1 through 255 on EBCDIC systems as the argument of the BYTE function and assign the result to a 1-byte character variable. The value of the character variable is the character whose position in the collating sequence is the number you used with the BYTE function. To produce the original number, use the character variable in the argument of the RANK function.

Using This Tip

This tip is useful either when storage space is extremely limited or when the SAS data set is extremely large. The advantage of this method over reading numbers into a character variable is that 1 byte can store two- and three-digit numbers, whereas standard character values require 1 byte per digit.

When you use this tip, remember that you may not be able to display the resulting character on your terminal or printer. To read the original number, you must decode the stored value with the RANK function. Also, the programming technique is not common and must be documented in the program to assist the programmer maintaining the program. Finally, both encoding and decoding the value require CPU time.

Acceptable

```
data large;
   length code $ 3;
   input code;
   cards;
108
109
110
;
```

More Efficient

```
data large2;
   length code2 $ 1;
   input code;
      /* encode */
   code2=byte(code);
   drop code;
   cards;
108
109
110
;

data temp;
   set large2;
      /* decode */
   code=rank(code2);
   drop code2;
run;
```

Chapter **8** Control Sorting

Understanding the Principle

Sorting observations—arranging them in order by values of one or more variables—is one of the most common operations performed with the SAS System. At the same time, sorting is relatively expensive in terms of CPU time and may require large amounts of work space.

Knowing when to sort data and how to sort them most efficiently is a simple way to save resources in your SAS programs.

The remainder of this chapter is a series of tips that illustrate specific ways to avoid unnecessary sorts and to sort data most efficiently in programs that do require sorting.

Note: When you invoke the SORT procedure, the computer system uses either a sorting module provided by SAS Institute, a sorting utility provided with the operating system, or a sorting utility provided by an independent vendor. The techniques shown in this chapter generally apply regardless of which sorting utility is used.

Tip 8.1: **Plan sorting to reduce the number of sorts.**

If appropriate, use two or more variables in the BY statement of one PROC SORT step rather than using two PROC SORT steps, each with one BY variable. If you need to create indexes for the data set for other reasons, you can create one composite index rather than multiple simple indexes. If several PROC steps require data sorted in the same order, group those PROC steps.

Using This Tip

Once sorted or indexed, observations remain in that order. Therefore, planning sorting allows you to reduce the amount of sorting in your programs in two ways: by sorting by more than one variable at once and by grouping DATA and PROC steps that require the same sorting order. Note that BY-group processing uses indexes if they are available but that it is not efficient to create an index solely for BY-group processing.

Group Once by These Variables

```
    /* Indicate these BY variables */
    /* in the SORT procedure      */
by a b c d;

    /* Indicate these key variables*/
    /* in the composite index      */
index create indxabcd=(a b c d);
```

Process These Statements

```
by a;
by a b;
by a b c;
by a b c d;
```

Sorting Before Each Procedure

```
     /* four sorts  */
     /* sort by sex */
proc sort data=test;
   by sex;
run;
proc print data=test;
   by sex;
run;

     /* sort by age */
proc sort data=test;
   by age;
run;
proc print data=test;
   by age;
run;

     /* sort by sex */
proc sort data=test;
   by sex;
run;
proc chart data=test;
   hbar score;
   by sex;
run;

     /* sort by age */
proc sort data=test;
   by age;
run;
proc chart data=test;
   hbar score;
   by sex;
run;
```

Grouping Procedures

```
     /* two sorts   */
     /* sort by sex */
proc sort data=test;
   by sex;
run;
proc print data=test;
   by sex;
run;
proc chart data=test;
   hbar score;
   by sex;
run;

     /* sort by age */
proc sort data=test;
   by age;
run;
proc print data=test;
   by age;
run;
proc chart data=test;
   hbar score;
   by age;
run;
```

Tip 8.2: Sort data only when necessary.

Know whether your work requires sorted data and the current order of the data before you use the SORT procedure.

Using This Tip

At times, you may sort data as a precaution. If you know whether the current step requires sorted data and whether the data are already sorted, you can avoid unnecessary sorts, thereby saving CPU time, memory, I/O operations, storage, and human resources.

You need sorted data in three cases:

□ when the DATA or PROC step uses a BY statement

□ when a procedure requires sorted data (without BY-group processing)

□ when you want sorted results for display (for example, an alphabetized list).

For BY-group processing, the SAS System checks the order of sorted data during execution of the DATA or PROC step. Any method that produces observations in the correct order is acceptable as preparation for BY-group processing. Possibilities include using the SORT procedure, using an existing index, and inputting the data in the correct order. (It is not efficient to create an index just for BY-group processing.)

Once in order, a data set remains in that order unless you alter the observations within it or move it to an operating system with a different collating sequence. There is no sort flag indicating that the data have been processed by the SORT procedure.

Some procedures, such as the MEANS, SUMMARY, and TABULATE procedures, can produce output in sorted order without using sorted input data, as discussed in "Tip 8.5: Use a CLASS statement in procedures." The SQL procedure can also produce a sorted output SAS data set without using sorted input by means of the ORDER BY clause in the SELECT and CREATE VIEW statements. The SQL procedure does not require sorted input SAS data sets.

Tip 8.3:   Sort as few observations and variables as possible.

Remove unneeded variables and observations either in a DATA step before the PROC SORT step or by using the DROP= or KEEP= data set option and the WHERE statement in the PROC SORT step.

Using This Tip

The SORT procedure consumes CPU time. In addition, it may create intermediate files that require I/O time. The fewer variables and observations the SORT procedure must move, the less CPU time and I/O time it uses. WHERE statements are especially efficient when the variables in the statement are indexed.

Acceptable

```
    /* OUT=LARGE2 is optional */
proc sort data=large out=large2;
   by id;
run;

data new;
   set large2(keep=id city state);
   where city in('Apex','Cary');
   more SAS statements
run;
```

More Efficient

```
    /* must use OUT=LARGE2    */
    /* to avoid altering LARGE */
proc sort data=large(keep=id city
     state) out=large2;
   where city in('Apex','Cary');
   by id;
run;

data new;
   set large2;
   more SAS statements
run;
```

Tip 8.4:	## Allow varying arrangements of observations within individual BY groups.

Specify the NOEQUALS option in the PROC SORT statement if it is not necessary for observations within BY groups to have the same relative order they had in the original SAS data set.

Using This Tip

If a BY group contains more than one observation, the SORT procedure by default outputs observations in the same order with respect to each other that they had in the original SAS data set. For example, suppose you sort a data set by a variable MONTH and the data set has three observations for January. The first observation for January also has the value 23 for another variable, SCORE; the second observation for January has the value 18 for SCORE; and the third observation has the value 45 for SCORE. The SAS System maintains the order of those observations in the output data set so that the values of SCORE are still ordered as 23, 18, and 45.

If you do not need the observations in that order, specify the NOEQUALS option to save CPU time and memory. Note that if a particular sorting utility does not support the NOEQUALS option, then the option is ignored; it produces no benefit.

Acceptable	**More Efficient**
| ```
proc sort data=big;
   by city;
run;
``` | ```
proc sort data=big noequals;
   by city;
run;
``` |

Tip 8.5:   Use a CLASS statement in procedures.

If a procedure supports both the CLASS statement and the BY statement, use the CLASS statement to eliminate the need for sorted input.

Using This Tip

The CLASS statement does not require data to be in sorted order, as the BY statement does. A procedure using the CLASS statement can process either sorted or unsorted data.

In particular, using the CLASS statement instead of the BY statement is helpful in the MEANS and SUMMARY procedures. In the TABULATE procedure, using a CLASS statement in the page dimension produces an effect similar to BY-group processing in other procedures. (You need a BY statement in the TABULATE procedure only if you want independent treatment of the BY groups or if you need to reduce memory consumption.)

The tradeoff is that CLASS statement processing uses more CPU time than does BY-group processing for a given procedure. However, it does not use as much as a separate PROC SORT step if the data are not already in the correct order for BY-group processing. In addition, a CLASS statement has no permanent effect on the input data set, whereas if you sort a data set, you can save the sorted version for later use.

Acceptable

```
proc sort data=new;
   by state;
run;

proc means data=new;
   var tv;
   by state;
run;
```

More Efficient

```
proc means data=new;
   var tv;
   class state;
run;
```

Tip 8.6:	**Mimic large sorts with other techniques.**

When a SAS data set is too large to sort on your computer system, use a DATA step with multiple data set names and OUTPUT statements to divide the original data set into smaller data sets, each of which contains observations with one value of the BY variable. You can also sort each of the resulting data sets by other BY variables. When all the individual data sets are in the correct order, concatenate them with a SET statement in another DATA step to produce a data set equivalent to sorting the original one.

Using This Tip

If a PROC SORT step executes successfully with your data, you do not need this tip. This tip does not save time or resources as measured in performance statistics. Instead, it allows you to perform work that otherwise would not be possible on a particular computer system. This tip allows you to substitute I/O operations and CPU time (by breaking up the original data set and processing the pieces separately) for unavailable memory.

Too Large to Run

```
libname in 'SAS-data-library-1';
libname out 'SAS-data-library-2';

proc sort data=in.huge
     out=out.huge2;
  by state city houshold;
run;
```

Uses Less Memory

```
libname in 'SAS-data-library-1';
libname out 'SAS-data-library-2';

   /* divide by STATE */
data state1 ... state50;
  set in.huge;
  select;
     when(state=1) output state1;
     more WHEN statements
     when(state=50) output
        state50;
     otherwise put _all_;
  end;
run;

   /* sort by other BY variables */
proc sort data=state1;
  by city houshold;
run;

more PROC SORT steps

proc sort data=state50;
  by city houshold;
run;

   /* Recombine           */
   /* Sorted by STATE, CITY, */
   /* and HOUSHOLD        */
data out.huge2;
  set state1 ... state50;
run;
```

Tip 8.7:   Use the most efficient sorting routine.

Use the SORTPGM= system option and related system options that may be required under your operating system to determine the best sorting routines for various situations.

Using This Tip

The SORTPGM= system option enables you to specify the sorting routine to be used by the SORT procedure. By default, the value of the SORTPGM= option is BEST. That is, the SAS System determines whether to use the sorting routine supplied by SAS Institute or a routine controlled by your operating system. In the majority of cases, the SAS System selects the most efficient routine. However, if your data sets are extremely large, you should investigate the various sorting routines available through your operating system to determine if a particular routine is best suited to your needs.

Some host operating systems require additional options in order to use the SORTPGM= option. See the SAS companion for your host operating system for details.

Chapter **9** Test Your Programs, Know Your Data

Understanding the Principle

Programs that fail due to preventable errors waste resources. Preventable program failures at critical times can be disastrous.

The best time to discover the shortcomings of your code or your data is during testing, where all of the elements of processing are under your control, and your programs and data can be easily repaired. Although testing is essential before you use programs in production, you gain efficiency by testing any programs you intend to use more than once.

Most of the techniques for finding and correcting the shortcomings of your code or your data fall into one of the following broad categories:

- using appropriate test data

- testing the right amount of code

- using the correct SAS tools to diagnose and fix problems

- controlling your test environment.

The tips in this chapter explain several practical techniques for finding and fixing flaws in your programs and your data.

Tip 9.1:   **Use realistic and complete test data.**

Use a subset of actual production data to test your programs. Use the OBS= SAS system option or data set option or a random selection algorithm to create a test SAS data set or a test raw data file. Make sure your test data contain every data condition your program recognizes before you use them for testing.

Using This Tip

When you test programs where sections of code conditionally execute based on the characteristics of the data, make sure that all possible conditions are present in your test data so that all of the code executes during testing. If your program uses existing data for input, test with a subset of the actual data rather than creating data yourself. Verify that any record layouts you have agree with the actual data.

With SAS data sets, use the OBS= SAS system option or data set option to extract a subset of data if it makes no difference whether the observations you choose for testing are contiguous. Use a random-number generating function (like RANUNI) to create a random subset of your data if contiguous observations won't provide all of the data characteristics you need for testing.

Acceptable

```
/* Create a test data set. */

data new;
   input x $ 1-10
         y 11-15 z $ 16-17;
   cards;
3/4" NUT   12345 Q
5/16" SCRW12346XV
more data lines
;
```

More Efficient

```
/* Create a test data    */
/* set from master file  */
/* with random record    */
/* selection, 10% sample. */

data new;
   set product.master;
   if ranuni(-1) le 0.10;
run;
```

Tip 9.2: **Develop and test incrementally.**

When coding a complicated application, individually test the pieces of code that make up the application. Test the entire application only after you are sure the individual components produce the right results.

Using This Tip

An application with many individual components is difficult to debug if you don't know which component causes the problem. If you test the individual components of a complicated application before you test the application as a whole, you are more likely to understand the source of any errors. After testing individual components, you understand the input and output requirements of the units of code that make up your program.

Acceptable

```
/* Test the entire program. */
proc contents data=&inds noprint
    out=temp (keep=name type);
run;

data _null_;
   set temp;
   if name="%upcase(&byvar)" then
      call symput('bvtype',type);
run;

proc sort;
   by state;
run;

data _null_;
   set sales end=eof;
   by state;
   if first.state then
      do;
   more SAS statements
run;
```

More Efficient

```
 /* Test this. */
proc contents data=&inds noprint
    out=temp (keep=name type);
run;

   /* Then test this. */
data _null_;
   set temp;
   if name="%upcase(&byvar)" then
      call symput('bvtype',type);
run;

   /* Then test this. */
proc sort;
   by state;
run;

   /* Then test this. */
data _null_;
   set sales end=eof;
   by state;
   if first.state then
      do;
   more SAS statements
run;
```

Tip 9.3:   **Request necessary messages.**

Specify the appropriate SAS system options to get all of the notes, messages, and warnings you need when you test code.

Using This Tip

For convenience, the SAS System enables you to turn various error messages or warnings on or off. You can also control the way your SAS session reacts if particular kinds of errors occur.

    The following list contains SAS system options that let you control your error and message processing:

DSNFERR	stops processing and prints a message if code contains a reference to a nonexistent SAS data set.
ERRORABEND	stops processing for errors that normally only generate a message.
ERRORS=	controls the maximum number of observations for which complete error messages are printed.
FMTERR	generates an error message when a format or informat associated with a variable cannot be found.
MERROR	generates a warning message when code contains a macro-like name that does not match a macro keyword or a user-defined macro.
MSGLEVEL=	allows control over the amount of information sent to the SAS log. Specify MSGLEVEL=I if you want to know whenever the SAS System uses a data set index.
NOTES	writes notes to the SAS log.
SERROR	generates a warning when a macro variable reference does not match any existing macro variables.
VNFERR	sets the error flag for missing variables when you use a _NULL_ data set.

Tip 9.4:   Reconcile log and output.

Always review the SAS log after submitting a job, even if your output looks correct and complete.

Using This Tip

The SAS System can provide reasonable-looking output even when there are problems with your syntax or with your data. The messages printed in the log can indicate problems with number of observations, missing values, uninitialized variables, values out of range, implicit type conversions, or other problems, but you must read the log to see them.

Make sure that all of the steps in a job stream complete normally by checking the log. The SAS System processes each step independently of every other step in a job stream. A step that stops because of errors in the middle of the job stream may not affect any other step in the stream or may cause misleading results. This is especially true with SAS procedures that provide summary statistics as output.

In the following example, the code on the left produces output, but the log on the right indicates problems with the code.

Produces Output

```
data perm.zoober;
   set perm.zoober;
   if _n_ lt 50 then do;
   accum=accum+_n_;
   zgroup=0;
run;


proc means data=perm.zoober
   mean css skewness;
run;
```

The Log

```
NOTE: The SAS System stopped
      processing this step
      because of errors.
WARNING: The data set
         PERM.ZOOBER may be
         incomplete. When
         this step was stopped
         there were 0 observations
         and 4 variables.
WARNING: Data set PERM.ZOOBER
         was not replaced because
         this step was stopped.
         -
      117
ERROR 117-185: There were 1
               unclosed  DO
               blocks.
```

Tip 9.5: Diagnose intermediate results.

In the DATA step, report values by placing PUT statements at various stages of the program to see if it is executing correctly. Use the PRINT procedure, the CONTENTS procedure, and the LIST statement to check the values of variables and the characteristics of SAS data sets or external files at step boundaries.

Using This Tip

When you test a complicated DATA step, the values of variables during execution signal whether your code is executing as you intend. If a DATA step executes without any error or warning messages, you should check any output SAS data sets or output files it creates to determine whether other DATA or PROC steps in your final job stream will get the input you intend.

Acceptable to Test	More Efficient to Test

```
data words;
   set words;
   by term;
   if first.term;
run;

data words;
   set words;
   if _n_ le 19 then
      do;
         c=_n_;
         col='A1 ';
      end;
   more SAS statements
run;
```

```
data words;
   set words;
   by term;
   if first.term;
run;

   /* Check data. */
proc print data=words;
run;

data words;
   set words;
   if _n_ le 19 then
      do;
         c=_n_;
         col='A1 ';
         /* Check values. */
         put 'For group A1'
             _n_= c= col=
             term=;
      end;
   more SAS statements
run;
```

Tip 9.6: Examine raw data before reading them.

Use the LIST statement in the DATA step to produce a dump of raw data records. If necessary, use file utilities to get a hexadecimal dump of some raw data records, and verify that the layout and contents of the data fields are in the format you expect.

Use system utilities or editors to clean up incorrect raw data. If necessary, change the INPUT statement to accommodate the differences.

Using This Tip

When you use an external file to create a SAS data set, verify that the file contains data in the form you expect. This precaution is especially important if you are unfamiliar with the file or if the file is very large.

In the following example, simply looking at the alignment of the data in the left column prevents data errors noted in the sample log in the right column.

Misaligned Data

```
020 356 ATAB 300 036 JSST
225 127 CXTR 410 023 ATAB
8 129 PQWW 312 300 CXTR
050 582 JSST 392 229 CBBT
more data lines
```

Resulting Log

```
1 data a97;
2    infile 'file-specification';
3    input id 1-3 lb 5-7
4        cd $ 9-12 qy 14-16
5        bg 18-20 c2 $ 22-25;
6 run;
     .
     .
     .
NOTE: Invalid data for ID in line 3 1-3.
NOTE: Invalid data for LB in line 3 5-7.
NOTE: Invalid data for QY in line 3 14-16.
NOTE: Invalid data for BG in line 3 18-20.
RULE: ----+----1----+----2----+----3----+---
3      8 129 PQWW 312 300 CXTR
ID=. LB=. CD=WW 3 QY=. BG=. C2=TR _ERROR_=1
_N_=3
```

Tip 9.7: Document your SAS data sets.

Use the CONTENTS statement in the DATASETS procedure to document several important characteristics of your SAS data sets. Use the OUT= option to produce an output SAS data set that you can read during a SAS job to find out data characteristics you need to control your processing.

Using This Tip

The CONTENTS statement in PROC DATASETS provides a wealth of information about your SAS data sets. The information you can get with this statement is a strong reason for keeping your data in SAS data sets.

The information available through using the CONTENTS statement in PROC DATASETS includes the following:

□ name of the SAS data set, including the libref used with the procedure

□ type of library member (DATA or VIEW)

□ engine used to read from and write to the data set

□ date the data set was created and date it was last modified

□ special data set type (such as CORR, COV, SSPC, EST, or FACTOR), if any

□ data set label, if one exists, and labels for each variable, if any

□ number of observations, variables, and indexes in the data set

□ length of the observations

□ number of observations marked for deletion (these are not included in the count of observations)

□ whether the data set is compressed

□ name, position, type, and length of each variable in the data set

□ starting position of each variable in the data set

□ format and informat for each variable in the data set, if any

□ name of each index, if any, and whether it requires unique values

□ whether observations with missing values for index variables are included in the index

□ list of variables making up each composite index

□ other information depending on your operating system.

Tip 9.8:

Label your variables and your SAS data sets.

Create labels for variables to provide more description than is possible in an eight-character variable name. Use the LABEL or ATTRIB statement in a DATA step to label variables in permanent SAS data sets. Label permanent SAS data sets with the LABEL= data set option. Use temporary variable labels in PROC steps to identify columns in procedure output reports.

Using This Tip

Labeling variables in SAS data sets provides more descriptive information about the values that each variable contains than it is possible to provide in the name of the variable. Using labels to include explanations of the values may enable you to avoid using formats and can also clarify reports you get as output from procedures.

Label data sets to indicate whether they are sorted, their sort order, whether they are subsets, what data sets they subset, who created them, and what purpose they serve. Include information to help you avoid doing unnecessary sorts, re-creating existing data sets unnecessarily, or using outdated or inappropriate information.

Acceptable

```
       CONTENTS PROCEDURE
Data Set Name: TDATA.RESPONSE
Member Type:   DATA
Engine:        V606
Created:       15:51 June 18, 1990
Last Modified: 15:51 June 18, 1990
Data Set Type:
Label:

#    Variable   Type Len   Pos
-------------------------------
3    FREQR1     Num   8     10
4    FREQR2     Num   8     18
5    FREQR3     Num   8     26
2    MAXFREQ    Num   8      2
1    TYPE       Char  2      0
```

More Efficient

```
       CONTENTS PROCEDURE
Data Set Name: TDATA.RESPONSE
Member Type:   DATA
Engine:        V606
Created:       15:51 June 18, 1990
Last Modified: 15:51 June 18, 1990
Data Set Type:
Label:    Sorted by TYPE on 17JUN90

# Variable Type Len Pos Label
-------------------------------
3 FREQR1   Num   8   10  MHz
4 FREQR2   Num   8   18  kHz
5 FREQR3   Num   8   26  MHz
2 MAXFREQ  Num   8    2  Max MHz
1 TYPE     Char  2    0  Device
```

Tip 9.9:	**Compile without executing to test syntax.**

Compile your programs without executing them to test for syntax errors and compile-time errors in one of the following ways:

□ Use the CANCEL option in the RUN statement.

□ Include a STOP statement in the main program to prevent execution.

□ Use the option OBS=0 in the DATA statement.

□ Structure a condition to ensure that the program doesn't execute (such as IF 0 THEN DO;).

□ Use the VALIDATE statement in the SQL procedure.

Using This Tip

When a DATA step in a job stream fails, the SAS System still executes later steps in the stream. If the point of the DATA step is to process information for later steps, you place data at risk when later processing occurs, even though the preceding DATA step fails.

Compiling a DATA step without executing it provides a quick way to catch the most common syntax and compile errors without putting any data at risk during testing. After you fix any syntax or compilation errors, your testing can then focus on other elements of the code.

Acceptable to Test	**More Efficient to Test**
`data test.payrol;`	`data test.payrol;`
`   set real.payrol;`	`   set real.payrol;`
`   wage40=wagehr-(wageot*1.5);`	`   wage40=wagehr-(wageot*1.5);`
`   more SAS statements`	`   more SAS statements`
`run;`	`run cancel;`

Tip 9.10:   Protect existing data when you test.

Use the NOREPLACE system option when you test using real SAS data sets so that experimental code doesn't accidentally overwrite important information. Use test libraries and data sets when you must test program output. Make copies of crucial data, or be aware of the backup plan at your site.

Using This Tip

When you test a program, specify the NOREPLACE option if the program refers to existing SAS data sets. Rely on intermediate results or internal diagnostics to judge whether the code works as you intend. You will still see if the code works, but you won't run the risk of overwriting legitimate data with experimental code.

When you need to test the output of your code, be sure the code specifies libraries and data sets set up specifically for testing. When you have no choice but to use real data, make backup copies of crucial information in case data-destroying errors occur.

Acceptable

```
data test.volatls;
   set real.gcmsdata;
   more SAS statements
run;
```

More Efficient

```
options noreplace;

data test.volatls;
   set real.gcmsdata;
   more SAS statements
run;
```

Tip 9.11: Use collapsing variables to group data.

Create a list of every possible combination of the classifying variables in a SAS data set. Develop a code that assigns a unique value to each different combination of all of the classifying variables. Create a format with the FORMAT procedure that translates the value of each code into a description of the combination it represents. Using a DATA step, add a variable to the main data set that contains the right code value for every observation; use your list of combinations to construct the logical conditions to do so. In the same DATA step, eliminate all of the other classifying variables from the data set.

Using This Tip

When you have a SAS data set that contains many variables that are just classification variables, you can keep all of the information that they represent by coding the combinations of possible values into a single variable, called a collapsing variable. After you store the meanings of the collapsing variable's values in a format, you can then eliminate all of the classifying variables, saving a considerable amount of storage space.

For example, suppose you have a data set with variables GENDER and DEPT. GENDER can have a value of M or F, and DEPT can have a value of DVLP, WRTG, or TSTG. Create a variable called COMBO, and for observations where GENDER is M and DEPT is DVLP, assign it the value 1. Where GENDER is M and DEPT is WRTG, assign COMBO the value 2, and so on, as in the following example. Then create a format that preserves the meanings of the values of COMBO, and eliminate GENDER and DEPT from the data set. This technique works with any number of classification variables.

By using this tip, you trade the processing time it takes to expand the collapsing variables for the disk space it takes to store all of the original variables.

Use This List

GENDER	DEPT	COMBO
M	DVLP	1
M	WRTG	2
M	TSTG	3
F	DVLP	4
F	WRTG	5
F	TSTG	6

Run This Program

```
proc format library=libref;
   value $gnjob
         '1'='MALE, DEVELOPMENT'
         '2'='MALE, WRITING'
         '3'='MALE, TESTING'
         '4'='FEMALE, DEVELOPMENT'
         '5'='FEMALE, WRITING'
         '6'='FEMALE, TESTING';
run;

data persnl.jobdesc;
   set persnl.jobdesc;
   length combo $ 1;
   drop gender dept;
   format combo $gnjob.;
   if gender='M' then do;
      if dept='DVLP'
         then combo='1';
      else if dept='WRTG'
           then combo='2';
      else combo='3';
   end;
   else do;
      if dept='DVLP'
         then combo='4';
      else if dept='TSTG'
           then combo='5';
      else combo='6';
   end;
run;
```

Chapter **10** Code Clearly

Understanding the Principle

An area of efficient programming that is often overlooked is the amount of time you as a programmer spend in interpreting programs. Because processing needs change, you will always need to modify or update your programs. The time you spend doing this depends on how clearly you write the programs initially.

Clearly written programs save time by allowing

- easier debugging during development

- easier maintenance

- easier modifications

- easier *transportability*, which is the ability to execute a program under an operating system other than the one under which you wrote it

- shorter learning curves

- fewer misused programs.

Most of the techniques for making your code clearer fall into one of the following broad categories:

- following conventions consistently

- using space to enhance clarity

- inserting explanations appropriately

- simplifying anything confusing.

The tips in this chapter describe several methods you can use to make your code more readable and easier to understand.

Tip 10.1: **Use a program header.**

Put a section at the beginning of your programs describing use, requirements, products, syntax, author, and modification history. Keep a file containing all header elements to use as a template for inserting headers into your programs.

Using This Tip

A program header serves as an introduction, overview, and usage guide to a program. As you become more experienced with the kind of maintenance your programs require, you can add items to your headers that you find useful.

To make using headers easier, keep a blank header in an external file to insert at the top of your programs. Then you can just fill in the blanks with the right information after you finish coding.

Acceptable

```
data temp.entries;
    infile file-specification;
    input word $ 1-15 def $ 16-75
        key $ 76-80;
run;

proc sort data=temp.entries;
    by key word;
    more SAS statements
run;
```

More Efficient

```
/**********************************/
/* PROGRAM: DICTNARY              */
/* USAGE: Creates a sorted        */
/*        list of new terms.      */
/* REQUIRES: Flat file output     */
/*           from INPUTX program. */
/* PRODUCES: SAS data set sorted  */
/*           by key and word.     */
/* AUTHOR: J. GALVAN, 4/17/91     */
/* UPDATED: 5/23/91-Added formats. */
/**********************************/
data temp.entries;
    infile file-specification;
    input word $ def $ 16-75
        key $ 76-80;
run;

proc sort data=temp.entries;
    by key word;
    more SAS statements
run;
```

Tip 10.2: Use comments in your programs.

Use comment statements before each program step. Explain any complicated logic, calculation, or obscure feature within your programs in comments. When you modify your programs, make sure to modify your comments.

Using This Tip

Just as a program header explains the use of the overall program, comments within a program explain its parts. In addition to announcing specific steps or specific parts of a step, you should take care to use comments anywhere that you think someone reading the code would benefit from clarification. For example, use comments when a logical condition directs processing, a complicated calculation occurs, a series of functions are nested within each other, or a little-used feature of the SAS language appears in your program.

Acceptable

```
options nodsnferr;

data _null_;
   call symput('tot',total);
   stop;
   set tempds nobs=total;
run;

options dsnferr;
```

More Efficient

```
/***************************/
/* Turn off error messages */
/* in case data set is     */
/* missing.                */
/***************************/

options nodsnferr;

/***************************/
/* Use a null DATA step    */
/* with SET to find number */
/* of obs and store in     */
/* macro variable.         */
/***************************/

data _null_;
   call symput('tot',total);
   stop;
   set tempds nobs=total;
run;

/***************************/
/* Turn missing data set   */
/* error messages back on. */
/***************************/

options dsnferr;
```

Tip 10.3: **Use meaningful variable names.**

Name your variables descriptively, so that each name has some connection with the contents or the function of each variable in your programs. Follow conventions across different programs so that you use variable names consistently.

Using This Tip

Interpreting the code for a complicated program becomes much easier if the names of the variables relate to their contents or function within a program. Suddenly, the program flow becomes more evident if you can tell simple index counters from running totals merely by looking at the variable names.

If you write or maintain several different programs, try to follow conventions in naming your variables in different programs. When you maintain several programs, it is less confusing if similarly named variables all represent the same thing in any program you must modify.

Acceptable	**More Efficient**

```
data x;                          data checks;
   retain x4 0;                     retain subtot 0;
   input x1 $ 1-10                  input payee $ 1-10
         x2 11-15                         deposit 11-15
         x3 16-20;                        payment 16-20;
   do x5=1 to 15;                   do counter=1 to 15;
   more SAS statements             more SAS statements
run;                             run;
```

Tip 10.4:   **Clearly separate steps in your programs.**

Use a RUN statement after every step. Use SKIP and PAGE statements liberally in your programs to isolate events in the SAS log.

Using This Tip

Use the RUN statement at the end of every step to ensure that the statistics and the code for that step are together in the SAS log. This makes it easier to match program steps with any warning or error messages. Also, the RUN statement provides an unambiguous boundary for each step, making your source code more readable.

Use SKIP and PAGE statements to emphasize key processing events in the SAS log even further with blank space. If your logs form part of your documentation, this provides valuable clarity. Within complicated DATA steps, you can make sure that particular parts of your programs appear at the top of a log page.

Acceptable

```
data temp;
   infile file-specification;
   input x y z;
proc sort data=temp;
options pagesize=55;
proc print data=temp;
proc plot data=temp;
   plot x*z;
   plot x*y;
more SAS statements
run;
```

More Efficient

```
data temp;
   infile file-specification;
   input x y z;
run;

page;
proc sort data=temp;
run;

skip 2;
options pagesize=55;
skip 2;
proc print data=temp;
run;

page;
proc plot data=temp;
   plot x*z;
   plot x*y;
more SAS statements
run;
```

Tip 10.5: Simplify complex expressions.

Reword excessively long or complicated expressions. Use parentheses to highlight and clarify the order of evaluation. Break complex conditions into several statements. Spread long expressions across several lines.

Using This Tip

When you need to use complicated calculations or conditions in a program, express them as simply as you can. Never combine expressions if you can achieve the same effect with individual statements. Even if you sacrifice processing efficiency, it will be minor compared with the human costs of maintaining unclear code. Use blank space and parentheses to make complicated expressions more readable.

Acceptable

```
SAS statements
xflag=htin>ftin*htout>ftout;
more SAS statements
```

More Efficient

```
SAS statements
if (htin gt ftin)
   and
   (htout gt ftout)
   then xflag=1;
else xflag=0;
more SAS statements
```

Tip 10.6: Group declarative statements.

Keep together statements that declare variables, set up arrays, or perform other housekeeping chores. Use a convention for placing declarative statements consistently across all of your programs.

Using This Tip

Declarative statements do their work during the compilation of your program. In addition to initializing variables or setting labels, they work like comments if you know where to look for them within your programs. By keeping LENGTH statements near the top of your programs, for example, you can always find out what variables your programs create, and how long they are, by scanning the first few lines.

Choose a convention for placing declarative statements that makes sense to you, and use it consistently in all of your programs. For example, one convention might be to let LENGTH, ARRAY, ATTRIB, and RETAIN statements precede *executable statements*, while DROP, KEEP, and LABEL statements follow executable statements. Executable statements do their work during program execution rather than during compilation.

Acceptable

```
data stats91.balance;
   infile file-specification;
   input a $ 1-10 b 11-15
         c $ 16-20;
   retain d 0 e 0.15
          f 'Account';
   if b lt 33 then g=e*b;
   label g 'Sales Tax';
   drop b;
   length h $ 20;
   h=f||a;
   more SAS statements
run;
```

Clearer

```
data stats91.balance;
   retain d 0 e 0.15
             f 'Account';
   length h $ 20;
   infile file-specification;
   input a $ 1-10 b 11-15
         c $ 16-20;
   if b lt 33 then g=e*b;
   h=f||a;
   more SAS statements
   label g 'Sales Tax';
   drop b;
run;
```

Tip 10.7: Use autocall macros instead of repetitive code.

Define macros to generate repetitive code in your SAS programs. Keep your macro definitions in an autocall library, and enable autocall processing with the MAUTOSOURCE system option. Substitute a macro invocation for the repetitive lines of code in your programs.

Using This Tip

Any time you write a SAS program that repetitively executes the same lines of code, you can put the code in a macro definition to make your code more compact and easier to read. By using a macro invocation instead of redundant code, you are less likely to make a typing error. Your code will also be easier to debug and modify, since you only need to make changes in one place.

Acceptable

```
data budg.genfund;
   infile file-specification;
   retain revb revytd
          expb expytd 0;
   input type $ moneyb moneyytd;
   select(type);
     when('A')
       do;
          name='Property';
          revb=revb+moneyb;
          revytd=revytd+
                 moneyytd;
       end;
     when('B')
       do;
          name='Sales';
          revb=revb+moneyb;
          revytd=revytd+
                 moneyytd;
       end;
     more SAS statements
   end;
run;
```

More Efficient

```
options sasautos=
        library-specification;

data budg.genfund;
   infile file-specification;
   retain revb revytd
          expb expytd 0;
   input type $ moneyb moneyytd;
   select(type);
     when('A')
       %revex(Property,
              revb,revytd);
     when('B')
       %revex(Sales,
              revb,revytd);
     more SAS statements
   end;
run;
```

Tip 10.8:   Put all code into your final program.

Put all code into your final program, rather than using %INCLUDE statements to retrieve code from external files. Use %INCLUDE statements during development so that you can easily change code fragments that appear in several places.

Using This Tip

Keeping %INCLUDE statements in your finished code makes it difficult for someone reading the code to follow program flow. Keeping all code within a program also reduces the number of I/O operations when you compile and execute your programs. Finally, by using code containing %INCLUDE statements, you risk referring to files that have been deleted or changed, putting your data at risk, and wasting resources on an aborted run.

When you work with volatile code, using %INCLUDE statements wisely can multiply your productivity. Using %INCLUDE statements to retrieve tested or already developed blocks of code is an important convenience when you develop or test a complicated job stream. Using %INCLUDE statements allows many individuals to work on the same job stream and simplifies the implementation of changes if you use particular %INCLUDE modules in many programs.

Acceptable	**More Efficient**
| ```
filename prox file-specification;

proc sort data=sales;
   by state;
run;
%include prox;
more SAS statements
``` | ```
proc sort data=sales;
   by state;
run;

proc freq data=sales;
   where state='NC';
   tables product*rep;
run;

proc means min max range maxdec=2;
   where state='NC';
   var gross;
run;
more SAS statements
``` |

Tip 10.9: Use arrays.

Define a series of variables as an array using the ARRAY statement. Refer to all of the elements of an array at one time using the asterisk (*) subscript operator rather than using variable lists.

Using This Tip

When you have variables in an array, refer to all of them at one time using the * subscript instead of typing out the full variable list that the array represents. This lets you develop shorter code with less chance of typographical errors without sacrificing program clarity.

Acceptable

```
data profit;
   infile file-specification;
   input carsls 5. applsls 5.
         clthsls 5. foodsls 5.
         lawnsls 5.;
   avgsales=mean(carsls,applsls,
                clthsls,foodsls,
                lawnsls);
   put carsls applsls clthsls
       foodsls lawnsls;
run;
```

More Efficient

```
data profit;
   infile file-specification;
   array sales(5) carsls applsls
                  clthsls foodsls
                  lawnsls;
   input sales(*) 5.;
   avgsales=mean(of sales(*));
   put sales(*);
run;
```

Tip 10.10:   Use straightforward logical conditions.

Express your logical conditions positively rather than negatively.

Using This Tip

When you need to express logical conditions within your programs, remember that conditions expressed positively are easier to understand than those expressed negatively. At the other extreme, do not contrive positive conditions when negative conditions are clearer.

Acceptable

```
data totals;
   set returns;
   if not (salary lt cuoff1)
      then rate=0.75;
   else rate=0.67;
more SAS statements
run;
```

Clearer

```
data totals;
   set returns;
   if salary ge cuoff1
      then rate=0.75;
   else rate=0.67;
more SAS statements
run;
```

Tip 10.11:   Always specify SAS data set names.

Always specify a SAS data set name in DATA and PROC statements within your programs.

Using This Tip

In complicated programs that create or use many SAS data sets or that have long blocks of code between step boundaries, using a data set name explicitly in each DATA or PROC statement makes it easy to see which data set each SAS step processes. It also reduces the chance that a later modification to the code will cause it to process the wrong data set.

Acceptable

```
data oldies contemp easy;
   set listener.all;
   if pref lt 70
      then output oldies;
   else if pref ge 85
         and pref ne 99
         then output contemp;
   else output easy;
run;

proc sort;
   by group;
run;

proc freq;
   where style="JAZZ";
   tables group*age;
run;
more DATA and PROC steps
```

More Efficient

```
data oldies contemp easy;
   set listener.all;
   if pref lt 70
      then output oldies;
   else if pref ge 85
         and pref ne 99
         then output contemp;
   else output easy;
run;

proc sort data=easy;
   by group;
run;

proc freq data=easy;
   where style="JAZZ";
   tables group*age;
run;
more DATA and PROC steps
```

Tip 10.12: Code for unknown data.

If you expect a variable to have *n* values, check for the $n+1^{th}$ value in a series of IF/THEN/ELSE statements or WHEN statements.

Using This Tip

If you program a list of conditional actions, you may miss one or more possibilities, or the data may have erroneous values other than those listed. If you don't specify an action for all remaining values, the program contains a logic error that is hard to spot. Executing a debugging statement such as the PUT statement for the $n+1^{th}$ condition identifies the observation in question.

Acceptable

```
if score=1 then action-1;
else if score=2 then action-2;
else if score=3 then action-3;
```

More Efficient

```
if score=1 then action-1;
else if score=2 then action-2;
else if score=3 then action-3;
else put '****** ' id= score= ;
```

Part 3
Summary of Tips

Part 3 contains two chapters that summarize programming tips.

Chapter 11 A Quick Look at Efficiency Tips

Use this chapter to find tips that save more than one resource or as an index to all tips. The following table lists tip numbers and names and indicates which resource each saves. Resources are identified by their icons.

Table 11.1   *Tips Summary Table*

▼ Tips	Resources Saved ▶	CPU	I/O	mem	📚	🖥️	🎖️
4.1	Read only the fields you need.	■	■		■		
4.2	Read selection fields first.						■
4.3	Store data in SAS data sets.	■	■				
4.4	Keep summaries of large SAS data sets.	■	■				
4.5	Store only the variables you need.				■		
4.6	Process only the variables you need.	■	■	■			
4.7	Create all data subsets at one time.	■	■				
4.8	Use informats for data transformations.	■				■	
4.9	Shorten data using formats and informats.				■		
4.10	Edit external files directly.	■					
4.11	Use one buffer for external file operations.	■	■	■			
4.12	Create indexes when appropriate.	■					
4.13	Use binary search, not null merge.		■	■			
5.1	Assign a value to a constant only once.	■					
5.2	Use constants in expressions.	■					
5.3	Condense constants in expressions.	■					
5.4	Use mutually exclusive conditions.	■					
5.5	Write conditions in order of descending probability.						■
5.6	Put missing values last in expressions.	■					
5.7	Check for missing values before using a variable in multiple statements.	■					
5.8	Put only statements affected by the loop in a loop.	■					
5.9	Assign many values in one statement.	■				■	

(continued)

Table 11.1   *(continued)*

▼ Tips	Resources Saved ▶	CPU	I/O	mem	⊜	⌗	⌾
5.10	Shorten expressions with functions.	■				■	
5.11	Edit character values with functions.	■				■	
5.12	Use the IN operator rather than logical OR operators.	■				■	
5.13	Use a series of conditions.	■					
5.14	Check for undesirable conditions and stop processing.					■	
5.15	Write the loop with the fewest iterations outermost.	■					
5.16	Use temporary arrays rather than creating and dropping variables.	■	■		■		
5.17	Use macros for repeated code.	■				■	
5.18	Create macro variables only when needed.	■					
5.19	Put a variable into only one array.	■					
5.20	Make array variables all retained or all unretained.	■					
5.21	Set the lower bound of arrays to 0.	■					
5.22	Use multidimensional explicit arrays.	■					
5.23	Use the Stored Program Facility.	■	■				
6.1	Let procedures do the work.					■	
6.2	Take advantage of output SAS data sets.						■
6.3	Use procedures to examine your data.					■	
6.4	Copy indexed SAS data sets with procedures.	■	■	■			
6.5	Store formats with the SAS data sets that use them.					■	
6.6	Use WHERE conditions in procedures.		■	■			
6.7	Use the SQL procedure to simplify your code.	■	■			■	
6.8	Take advantage of column operations in the SQL procedure.	■	■				
6.9	Get cross-tabulations with output SAS data sets.	■	■				
7.1	Reduce the storage space for variables.				■		
7.2	Shorten SAS date values.				■		

(continued)

Table 11.1   (*continued*)

▼ Tips Resources Saved ▶	CPU	I/O	mem	▤	▦	▧
7.3 Know the rules for creating null data sets.	■	■		■		
7.4 Avoid default type conversions.	■					
7.5 Use character rather than numeric variables.				■		
7.6 Create separate variables instead of repeating type conversions.	■					
7.7 Use SAS configuration and autoexec files.					■	
7.8 Compress large SAS data sets.		■		■		
7.9 Eliminate the macro facility if your programs do not need it.	■					
7.10 Store numbers as 1-byte character values.				■		
8.1 Plan sorting to reduce the number of sorts.	■					
8.2 Sort data only when necessary.						■
8.3 Sort as few observations and variables as possible.						■
8.4 Allow varying arrangements of observations within individual BY groups.	■		■			
8.5 Use a CLASS statement in procedures.	■					
8.6 Mimic large sorts with other techniques.			■			
8.7 Use the most efficient sorting routine.	■					
9.1 Use realistic and complete test data.					■	
9.2 Develop and test incrementally.					■	
9.3 Request necessary messages.					■	
9.4 Reconcile log and output.					■	
9.5 Diagnose intermediate results.					■	
9.6 Examine raw data before reading them.						■
9.7 Document your SAS data sets.					■	
9.8 Label your variables and your SAS data sets.					■	
9.9 Compile without executing to test syntax.					■	
9.10 Protect existing data when you test.					■	
9.11 Use collapsing variables to group data.				■		
10.1 Use a program header.					■	

(*continued*)

Table 11.1   *(continued)*

▼ Tips	Resources Saved ▶	CPU	I/O	mem	(disk)	(program)	(award)
10.2	Use comments in your programs.					■	
10.3	Use meaningful variable names.					■	
10.4	Clearly separate steps in your programs.					■	
10.5	Simplify complex expressions.					■	
10.6	Group declarative statements.					■	
10.7	Use autocall macros instead of repetitive code.					■	
10.8	Put all code into your final program.					■	
10.9	Use arrays.					■	
10.10	Use straightforward logical conditions.					■	
10.11	Always specify SAS data set names.						■
10.12	Code for unknown data.					■	

Chapter 12 Summary of Tips By Resource

Use this chapter to find all the tips that save a resource. There are six lists, one per resource, and the tips are arranged by tip number within the lists.

Tips That Are Always Useful

4.2	Read selection fields first.
5.5	Write conditions in order of descending probability.
6.2	Take advantage of output SAS data sets.
8.2	Sort data only when necessary.
8.3	Sort as few observations and variables as possible.
9.6	Examine raw data before reading them.
10.11	Always specify SAS data set names.

Tips That Save Memory

4.6	Process only the variables you need.
4.11	Use one buffer for external file operations.
4.13	Use binary search, not null merge.
6.4	Copy indexed SAS data sets with procedures.
6.6	Use WHERE conditions in procedures.
8.4	Allow varying arrangements of observations within individual BY groups.
8.6	Mimic large sorts with other techniques.

Tips That Save CPU Time

CPU

4.1	Read only the fields you need.
4.3	Store data in SAS data sets.
4.4	Keep summaries of large SAS data sets.
4.6	Process only the variables you need.
4.7	Create all data subsets at one time.
4.8	Use informats for data transformations.
4.10	Edit external files directly.
4.11	Use one buffer for external file operations.
4.12	Create indexes when appropriate.
5.1	Assign a value to a constant only once.
5.2	Use constants in expressions.
5.3	Condense constants in expressions.
5.4	Use mutually exclusive conditions.
5.6	Put missing values last in expressions.
5.7	Check for missing values before using a variable in multiple statements.
5.8	Put only statements affected by the loop in a loop.
5.9	Assign many values in one statement.
5.10	Shorten expressions with functions.
5.11	Edit character values with functions.
5.12	Use the IN operator rather than logical OR operators.
5.13	Use a series of conditions.
5.15	Write the loop with the fewest iterations outermost.
5.16	Use temporary arrays rather than creating and dropping variables.
5.17	Use macros for repeated code.
5.18	Create macro variables only when needed.
5.19	Put a variable into only one array.
5.20	Make array variables all retained or all unretained.
5.21	Set the lower bound of arrays to 0.
5.22	Use multidimensional explicit arrays.
5.23	Use the Stored Program Facility.
6.4	Copy indexed SAS data sets with procedures.
6.7	Use the SQL procedure to simplify your code.
6.8	Take advantage of column operations in the SQL procedure.

6.9	Get cross-tablulations with output SAS data sets.
7.3	Know the rules for creating null data sets.
7.4	Avoid default type conversions.
7.6	Create separate variables instead of repeating type conversions.
7.9	Eliminate the macro facility if your programs do not need it.
8.1	Plan sorting to reduce the number of sorts.
8.4	Allow varying arrangements of observations within individual BY groups.
8.5	Use a CLASS statement in procedures.
8.7	Use the most efficient sorting routine.

Tips That Save I/O Operations

4.1	Read only the fields you need.
4.3	Store data in SAS data sets.
4.4	Keep summaries of large SAS data sets.
4.6	Process only the variables you need.
4.7	Create all data subsets at one time.
4.11	Use one buffer for external file operations.
4.13	Use binary search, not null merge.
5.16	Use temporary arrays rather than creating and dropping variables.
5.23	Use the Stored Program Facility.
6.4	Copy indexed SAS data sets with procedures.
6.6	Use WHERE conditions in procedures.
6.7	Use the SQL procedure to simplify your code.
6.8	Take advantage of column operations in the SQL procedure.
6.9	Get cross-tablulations with output SAS data sets.
7.3	Know the rules for creating null data sets.
7.8	Compress large SAS data sets.

Tips That Save Storage

4.1	Read only the fields you need.
4.5	Store only the variables you need.
4.9	Shorten data using formats and informats.
5.16	Use temporary arrays rather than creating and dropping variables.
7.1	Reduce the storage space for variables.
7.2	Shorten SAS date values.
7.3	Know the rules for creating null data sets.
7.5	Use character rather than numeric variables.
7.8	Compress large SAS data sets.
7.10	Store numbers as 1-byte character values.
9.11	Use collapsing variables to group data.

Tips That Save Programmer Time

4.8	Use informats for data transformations.
5.9	Assign many values in one statement.
5.10	Shorten expressions with functions.
5.11	Edit character values with functions.
5.12	Use the IN operator rather than logical OR operators.
5.14	Check for undesirable conditions and stop processing.
5.17	Use macros for repeated code.
6.1	Let procedures do the work.
6.3	Use procedures to examine your data.
6.5	Store formats with the SAS data sets that use them.
6.7	Use the SQL procedure to simplify your code.
7.7	Use SAS configuration and autoexec files.
9.1	Use realistic and complete test data.
9.2	Develop and test incrementally.
9.3	Request necessary messages.
9.4	Reconcile log and output.
9.5	Diagnose intermediate results.
9.7	Document your SAS data sets.
9.8	Label your variables and your SAS data sets.
9.9	Compile without executing to test syntax.

9.10   Protect existing data when you test.

10.1   Use a program header.

10.2   Use comments in your programs.

10.3   Use meaningful variable names.

10.4   Clearly separate steps in your programs.

10.5   Simplify complex expressions.

10.6   Group declarative statements.

10.7   Use autocall macros instead of repetitive code.

10.8   Put all code into your final program.

10.9   Use arrays.

10.10   Use straightforward logical conditions.

10.12   Code for unknown data.

Part 4
Appendix

Appendix   **Generating Data for Testing Program Efficiency**

Appendix # Generating Data for Testing Program Efficiency

Introduction

This appendix provides some techniques for generating data in external files or in SAS data sets. You can use the routines in the following sections to generate data to use in benchmarking your programs for efficiency.

Generating Data: Sample Programs

The following sections show two techniques for generating data. The first technique uses several nested loops. The second technique uses macros, functions, and call routines to generate random numeric and character data.

Generating Data with Nested Loops

You can generate a SAS data set or an external file containing as many observations as you need by using a program that contains nested loops. Using this technique, you let the changing value of several index variables provide different observations in a data set or an external file. In the following example, executing the code as it is shown produces an external file with 250,000 records and a SAS data set with 250,000 observations. You can vary the number of records or observations you generate by changing the number of times each loop iterates.

If you only want to create an external file with the following program, eliminate the LIBNAME and OUTPUT statements and use _NULL_ for the data set name in the DATA statement. If you only want to create a SAS data set, eliminate the FILENAME, FILE, and PUT statements.

The following program uses a nested loop technique to create both a SAS data set and an external file:

```
libname eff 'library-specification';
filename rawdata 'file-specification';
data eff.sasdata(drop=i);
   file rawdata;
   do test=1 to 4;
      do status=1 to 5;
         do flag=1 to 10;
            do code=1 to 5;
               do i=1 to 10;
                  value=uniform(0);
                  do j=1 to 25;
                     random=uniform=(0);
                     output;
                     put a1 test a10 status
                         a20 flag a30 code
                         a40 value a60 random;
                  end;
               end;
            end;
         end;
      end;
   end;
   stop;
run;
```

Generating Random Numeric and Character Data

If you need to test programs with data containing unpredictable patterns, you can use SAS functions, call routines, and formats to create random numeric and character data.

Using the techniques shown in the following example, you use a macro to generate a random number with magnitude and rounding characteristics you specify, and construct character values based on a bounded value probability distribution. In the following example, executing the code as it is shown produces an external file with 25,000 records and a SAS data set with 25,000 observations. You can vary the number of records or observations you generate by changing the number of times the loop iterates.

If you only want to create an external file with the following program, eliminate the LIBNAME and OUTPUT statements and use _NULL_ for the data set name in the DATA statement. If you only want to create a SAS data set, eliminate the FILENAME, FILE, and PUT statements.

The following program uses a macro containing the RANUNI function, a user-defined format, and the CALL RANTBL routine to generate both random character and random numeric data.

```
%macro gen(mult,round);
   round((ranuni(-1)*&mult),&round)
%mend gen;
proc format;
   value nlet 1='A'  2='B'  3='C'
              4='D'  5='E'  6='F'
              7='G'  8='H'  9='I'
             10='J' 11='K' 12='L'
             13='M' 14='N' 15='O'
             16='P' 17='Q' 18='R'
             19='S' 20='T' 21='U'
             22='V' 23='W' 24='X'
             25='Y' 26='Z';
run;


libname eff 'library-specification';
filename rawdata 'file-specification';

data eff.sasdata;
   file rawdata;
   retain p1-p26 0.0384615 j1-j9 0.1111111 seed -1;
   drop p1-p26 j1-j9 seed i x1-x2;
   do i=1 to 25000;
      call rantbl(seed,of j1-j9,jday);
      call rantbl(seed,of p1-p26,x1);
      call rantbl(seed,of p1-p26,x2);
      rcode=put(x1,nlet.)||put(x2,nlet.);
      sales=%gen(1000,1);
      exps=%gen(1000,1);
      salhrs=%gen(1000,1);
      exphrs=%gen(1000,1);
      invin=%gen(1000,1);
      invout=%gen(1000,1);
      output;
      put @1 jday @3 rcode @6 sales @15 exps @22 salhrs
          @30 exphrs @36 invin @42 invout;
   end;
run;
```

Glossary

benchmarking
the process of measuring the difference in resources used by comparable programming techniques.

buffer
a temporary storage area reserved for holding data after they are read or before they are written.

compilation
the automatic translation of SAS statements into machine code.

CPU
an abbreviation for Central Processing Unit. The CPU is the primary hardware unit of a computer system consisting of storage elements (registers), computational circuits (arithmetic-logic units), control circuits, and I/O ports.

CPU time
the time the CPU spends in performing the calculations or other operations you request.

data reduction
the transformation of data into corrected, ordered, and simplified form.

data storage
the amount of space on disk or tape required to keep your data.

data transformation
the changing of a data value from one form to another (for example, a calendar date to a SAS date value or vice versa).

default type conversion
a numeric-to-character or character-to-numeric conversion that the SAS System performs automatically.

efficiency
the ability to obtain more results from fewer human or computer resources.

elapsed time
the amount of clock time needed to receive the result of a computer program, including time used by the computer to execute your program and the time your program spends waiting for resources (such as memory or a printer). Elapsed time is also known as throughput.

executable statement
a SAS statement that is completed after compilation and one that can be executed on an individual observation in a DATA step.

human efficiency
the amount of programming time required to develop and to maintain a program.

I/O time
an abbreviation for input/output time. I/O time is the time the computer spends on moving data from storage areas, such as disk or tape, into memory for work (input time) and moving the result out of memory to storage or to a display device, such as a terminal or a printer (output time).

main memory
an area that can be accessed rapidly by the CPU and from which instructions are executed and data are operated on.

memory
(1) an area into which a unit of information can be copied, which will hold the information and from which the information can be obtained at a later time. (2) the size of the work area that the CPU must devote to the operations in a program.

overhead
in benchmarking, the additional resources used to move a component of the SAS System (such as a procedure) into main memory the first time a program uses that component.

page
a portion of a SAS data set that the SAS System moves between external storage and memory in one input/output operation.

performance statistics
the data that measure the amount of resources a program uses when it compiles and executes.

program compilation
See compilation.

step boundary
a point in a SAS program at which the SAS System recognizes that a DATA or PROC step is complete. It consists of either a RUN statement; a QUIT statement in an interactive procedure; a semicolon following data lines; the word DATA or PROC indicating the beginning of the next step; an ENDSAS statement; or the end of a noninteractive program.

storage
See data storage.

tradeoff
the process of allowing a program to use more of one resource in order to decrease the use of another resource.

throughput
See elapsed time.

transportability

the ability of a program to execute under an operating system other than the one under which it was written.

wildcard

a character used to represent possible characters at a particular position in a word in order to generalize the word.

Index

Your Turn

If you have comments or suggestions about base SAS software or *SAS Programming Tips: A Guide to Efficient SAS Processing*, please send them to us on a photocopy of this page.

Please return the photocopy to the Publications Division (for comments about this book) or the Technical Support Division (for suggestions about the software) at SAS Institute Inc., SAS Campus Drive, Cary, NC 27513.